Quantitative Pharmacology

An Introduction to Integrative Pharmacokinetic-Pharmacodynamic Analysis

Quantitative Pharmacology
An Introduction to Integrative Pharmacokinetic-Pharmacodynamic Analysis

by

Johan Gabrielsson & Stephan Hjorth

Faculty of Veterinary Medicine and Animal Sciences
Swedish University of Agricultural Sciences SLU,
Uppsala, Sweden
AstraZeneca R&D, Mölndal, Sweden

APOTEKARSOCIETETEN
SWEDISH ACADEMY OF PHARMACEUTICAL SCIENCES

This book is dedicated to our families, Barbro and Jenny Gabrielsson, and Anette, Linn, Sanna and Lowe Hjorth, without the support of which it might not have come to fruition.

Quantitative Pharmacology An Introduction to Integrative Pharmacokinetic-Pharmacodynamic Analysis, by Johan Gabrielsson & Stephan Hjorth

Cover picture: Painting by J. Gabrielsson

www.apotekarsocieteten.se
Webshop: www.lma-books.se

ISBN: 978-91-979452-3-3
Printografen AB, Sweden 2012

Foreword

The aim of this book has been not only to introduce, review and discuss pharmacokinetic paradigms useful in drug development, but also to incorporate these into various pharmacological and physiological contexts, in the hope of extending the overall understanding of how an integrated pharmacokinetic-pharmacodynamic approach helps optimizing therapeutic principles towards man.

By providing adequate tools, methods and approaches, we have tried to disentangle and shed light on some of the complex interactions that govern PKPD relationships. We have endeavored to illustrate in selected case scenarios and examples, common pitfalls and solutions to challenges encountered in the pharmacokinetic evaluation of compounds. Throughout, we have strived to point out aspects that deserve particular attention when analyzing PK data in relation to PD response, and also suggested ways to optimize the design of such studies to provide a solid basis for decision-making.

Through this, we hope that the book has helped furthering the understanding of the subject matter, and that the reader will have found it worthwhile, including finding use of it also as a reference to assist with experimental design and analysis of PKPD studies onwards.

Ultimately, we hope that the contents of the book will encourage experimenters to try a more holistic approach to drug development, through which compounds may be evaluated in a more efficient manner, with positive repercussions both on quality and cost towards future drug development.

This book is dedicated to our families, without the support of which it might not have come to fruition. Also, over the years we have had countless interactions on PK/PD-related matters across Bioscience and Pharmacokinetics discipline functions. These exchanges have provided several educational examples (data adapted for proprietary reasons) demonstrating the advantages of adopting an integrative PK/PD approach to drug development. Our thanks are therefore also due to all colleagues and friends in- and outside of AstraZeneca that have contributed to widening our experiences in these contexts. This has helped identifying vital requirements and shaping our views on how to accomplish those for the presumptive audience to this book. The Support from Apotekarsocieteten – Swedish Academy of Pharmaceutical Sciences and Printografen AB are also appreciated.

Johan Gabrielsson
Professor of Integrative Pharmacology
Department of Biomedicine and Veterinary Public Health
Faculty of Veterinary Medicine and Animal Sciences
Swedish University of Agricultural Sciences
SE-750 07 Uppsala
Sweden
Johan.Gabrielsson@SLU.SE
Tel: +46-(0)18-67 31 72

Biography

Johan Gabrielsson has been Senior Principal Scientist at AstraZeneca R&D Mölndal. His responsibilities included kinetic/dynamic related project tasks in the CV & GI & CNS & Cancer & Biologics areas. He is author of the textbook *'Pharmacokinetic and Pharmacodynamic Data Analysis: Concepts and Applications'* 4th ed. (2006). He is professor of Integrative Pharmacology at the Swedish Agricultural University in Uppsala, Sweden. He is also academically affiliated with Dept. of Pharmaceutics, Univ. Tennessee and University of Manchester. He has published extensively in the field of pharmacokinetic-pharmacodynamic modelling and reasoning, and has run numerous courses internally and externally in the area of biological data analysis since 1985 in Europe, the US and Asia (>4000 participants) at both the undergraduate and graduate level. His research focuses on modelling different aspects of endogenous turnover, such as functional tolerance and rebound phenomena, physiological limits and target-mediated drug disposition in collaboration with Professor LA Peletier, Leiden University. He has been external examiner on several PhD Theses abroad.

Stephan Hjorth
PhD, professor
Principal Scientist
Department of Bioscience
AstraZeneca R&D
SE-431 83 Mölndal
Sweden
Stephan.Hjorth@AstraZeneca.com
Tel: +46-(0)31-776 1000

Biography

Stephan Hjorth is Principal Scientist at the Bioscience Dept., AstraZeneca R&D, Mölndal. His main responsibilities includes pharmacodynamic-related project tasks in the CV&GI, metabolism, area. Before joining AstraZeneca R&D he spent ~25 years in various positions at the Dept. of Pharmacology, Gothenburg Univ., and latterly as appointed professor at the Dept. of Pharmacology & Clinical Neurosciences, Umeå Univ., doing research and teaching at the undergraduate as well as the graduate level. He has published extensively, mainly in the field of preclinical neuro- and psychopharmacology, and also acted as editor for the Journal of Neural Transmission. He is a member of several national and international scientific societies, has reviewed grant applications to the EU Framework programme and other foreign grant agencies, and been external examiner at several PhD dissertations in Sweden and abroad. His current research primarily focuses on discovery and development of drugs for obesity and appetite control, including efforts towards animal models useful in translational modelling of PK-PD relationships.

Table of Contents

Chapter 1
Introduction - What is PK/PD and Quantitative Pharmacology?

Preface

Pharmacokinetics (PK) and *pharmacodynamics* (PD) comprise traditionally distinct disciplines within *pharmacology*, the study of the interaction of drugs with the body. Whereas pharmacokinetics looks at how the body handles drugs, pharmacodynamics looks at the effects of drugs on biological systems. It is because of the traditional interests of pharmacokineticists and pharmacologists (who investigate pharmacodynamics) that they may frequently work more or less in isolation of each other. However, this sort of approach greatly restricts the detailed understanding of both known and potentially novel drugs.

It is our intention in this book to show that by adopting a different approach – one in which the activities of these two disciplines are deliberately, intimately and systematically integrated – our understanding of drugs and the efficiency and effectiveness of drug discovery and development will be greatly enhanced. Failure to take the collected PK and PD information into account may lead to prematurely discounting drug candidates with great potential, or promoting agents which have inferior characteristics. Either of these scenarios are of course undesirable. Not only may they result in substandard exploitation of experimental data – whereby valuable critical info to a project is lost, with potential impact on development timelines and associated costs – but also because the potential for discovering and bringing to man new and improved therapeutical principles may be seriously eroded.

Because knowledge about drugs in all its guises benefits from an understanding of their joint pharmacokinetic and pharmacodynamic properties, the book is written with a broad audience in mind. It includes pharmacologists of all sorts (traditional, safety, and clinical), pharmacokineticists, medicinal chemists, clinicians, statisticians, and project leaders, to name some obvious groups. With such a diverse audience in mind, every effort has been made to build up concepts rather than to take things for granted. This said, however, some basic knowledge in the areas of physiology, pharmacology, and perhaps pharmacokinetics would be advantageous for readers. Although the approach to experimental design and analysis outlined has implications for medicinal and formulation chemists, the book does not discuss the intricacies of drug structure or formulation. Thus, although a knowledge of chemistry would be beneficial, it is certainly not a pre-requisite. Finally, because much of pharmacokinetics is directed towards mathematically modelling drug behaviour, it is inevitable that a book like this will include mathematical concepts and equations. However, only rudimentary knowledge in this regard is required in order to appreciate the consequences of the approach promoted (– certainly, the intention is not to turn readers into mathematicians).

This chapter

- introduces the basics of pharmacokinetics and pharmacodynamics, and explains what is meant by quantitative pharmacology or integrative pharmacokinetics-pharmacodynamics (PK/PD)
- outlines the implications of integrating pharmacokinetic and pharmacodynamic analysis, and lays the groundwork to show the potential consequences of failing to do so. In so doing, it points to the importance of PK/PD for effective and efficient drug discovery and development.
- Finally, it introduces the principles behind using different types of biomarkers which enables better interpretation of the preclinical and clinical features of a drug, and facilitates predictions of the clinical outcome of treatment in man.

The concepts introduced in this chapter are developed further in subsequent chapters.

Wherever possible, data from known drugs rather than from make-believe drug profiles are presented. However, if drugs are not identified by name it is because the information remains confidential. Finally, a brief but generic Bibliography is included to provide interested readers with sources from which to obtain more detailed information.

1.1 The scope of pharmacokinetics and pharmacodynamics

1.1.1 Pharmacokinetics

Before it can act, a drug must reach the site or sites at which it is able to exert an effect. Depending on how it is administered or taken, this may mean that the drug has to move through different parts of the body. Once taken or absorbed, and distributed, it is usually changed chemically (degraded or metabolised), before being finally removed from the body. The branch of pharmacology that examines these processes of what the body does to a drug is *pharmacokinetics*. In other words, pharmacokinetics considers drug *absorption*, *distribution*, *metabolism*, *removal* (*elimination*) and *turnover*.

Drugs are administered in a variety of ways. For example, they may be

- injected (rapid administration) or infused (administration over a longer time frame) directly into the blood (usually intravenous), cerebrospinal fluid, a muscle, or under the skin
- inhaled by mouth into the lungs
- given as drops or spray into the nose
- delivered directly to the gastro-intestinal tract by oral ingestion or rectal administration
- applied topically particularly to the skin, cornea of the eye, lips, lining of the mouth, nose, vagina or rectum.

While the intended site of action in part determines a drug's route of administration, its chemical and physical properties are also taken into consideration. In most cases the drug must be absorbed into the blood stream before it can be distributed to its site of action, metabolised, and eliminated. Clearly, all of these processes take time. In addition, for a drug to exert an effect it must not only be given in adequate amounts and in an appropriate formulation, but it must reach its site of action in adequate amounts (concentrations). Thus, pharmacokinetic assessment involves determining the time-course of the concentration of the drug (and, if relevant, its metabolite/s) in different body spaces or compartments (plasma, blood, urine, cerebrospinal fluid, other tissues), and attempting to relate this to the time-course of drug action.

1.1.2 Pharmacodynamics

Most drugs act by interacting with some biological target to produce an effect. This target may be on the cell-membrane, or it may be inside the cell or its nucleus. It may also be an enzyme, or some other cellular or sub-cellular structure. By acting at specific target sites in different parts of the body, different drugs produce their own specific effects. It is the study of the biochemical and physiological effects of drugs, and of their mechanisms of action, that comprise the branch of pharmacology known as *pharmacodynamics*. In other words, whereas pharmacokinetics looks at what the body does to a drug, pharmacodynamics looks at what a drug does to the body.

To be able to produce an effect, a drug must reach its target not only at an adequate concentration, but also for an appropriate period of time. Thus, the magnitude and time-course of a drug-induced response depends not only on the particular drug given, but also on the amount given, and its route and mode of administration. Furthermore, the nature of the response depends on the type and location of target(s) with which the drug interacts. Even if a drug is very specific for a particular target, it is more often the rule rather than the exception that the very target in question is found in more than one location in the body (for example, in brain as well as in peripheral tissue). It is thus no surprise that most drugs produce more than one effect. While the desired effect of the drug is by definition beneficial, some effects are not wanted, and some may even be harmful. These unwanted effects are termed adverse drug reactions or side effects. Just as it is important to know that a drug produces its desired effect (*i.e.*, that it is efficacious) at a particular dose, so it is important to know what side effects it produces, and at what doses. It is only when armed with this information that potential toxicity can be prevented or minimized.

1.2 The basic tools of pharmacokinetics and pharmacodynamics

As suggested above, a number of factors determine whether or not the concentration of drug at its site of action is optimal. These include the drug's chemical and physical properties, its formulation and dose, and the route, speed and frequency of administration. Just as there are marked differences in the rate of absorption of different drugs, and in the extent to which they are absorbed, there is variability in the rate at which drugs are distributed throughout the body, in the tissues to which they are distributed, and in the nature and site of their metabolism and elimination. In addition, drugs do not behave identically in different individuals (or species), or even in any one person at different times. Whereas pharmacokinetics considers the how concentration, time-course and removal of a drug varies in different bodily compartments, pharmacodynamics looks at the drug's effects and mechanisms of action by considering the magnitude and time-course of the responses following a particular dose, route and mode of administration. Not surprisingly, these two branches of pharmacology use different tools to achieve these ends.

The types of responses investigated in pharmacological studies are many and varied. They include changes in sensation and body movement; visualization of drug uptake and structural changes; receptor binding, biochemical, genetic, immunological and microbiological assays; changes in enzyme-mediated responses, body temperature, cardiovascular and respiratory parameters, and many more. They also include direct responses and surrogate (or substitute) markers, as well as *in vitro* and *in vivo* (respectively, in a test tube or artifical system, and in a living system) techniques. In short, pharmacology draws on many different branches of the biological sciences and uses many different techniques. However, regardless of the response measured or the technique used, the major feature of traditional pharmacodynamic studies is to determine the relationship between the magnitude (intensity) of a response R and its time-course t, and between the magnitude of the response R and concentration C (or dose D) of drug.

Also in pharmacokinetic studies it is the concentration of drug in different body compartments (*e.g.*, plasma, urine, tissue) after a given dose, that is related to the time-course of a pharmacological response. Because of the large number of factors influencing a drug once it has been administered, pharmacokinetics makes use of numerous measures to obtain information about drug absorption, distribution, availability, metabolism and excretion, and it develops mathematical models to explain the relationships between these measures. To name just a few, common measures employed in pharmacokinetics apart from drug concentration include

- clearance Cl (a measure of the body's capacity to remove a drug)
- volume of distribution V (the apparent volume into which a drug distributes after dosing)

- bioavailability F (the fraction of an extravascularly administered dose of drug that reaches the systemic circulation)
- and the independent variable time t.

Some measures used in pharmacokinetic modelling (and, of course, in pharmacodynamics) are constants (*e.g.*, dose), some are variables (*e.g.*, concentration, response, time), and some are parameters to be estimated by fitting models to the data (*e.g.*, clearance, volume of distribution, potency).

1.3 Quantitative pharmacology

1.3.1 Basic principles

In the course of investigating drug absorption, distribution, and elimination, pharmacokinetics also takes into account aspects of physiology such as turnover (production and loss of endogenous compounds, including, for example, the natural production and breakdown of hormones or proton pumps) and binding (the interaction of a drug with other molecules such as plasma proteins). Because all of the above-mentioned aspects of pharmacokinetics affect drug concentration, it is their combined influence which ultimately determines the concentration of drug at the target sites, and therefore available to exert a pharmacodynamic effect. In other words, as shown schematically in Figure 1.1, pharmacokinetics is inseparable from the primary efficacy (ability of a drug to produce its desired effect) and secondary or toxicity effects (ability of a drug to produce an undesired effect) of a drug, and therefore from its pharmacodynamics.

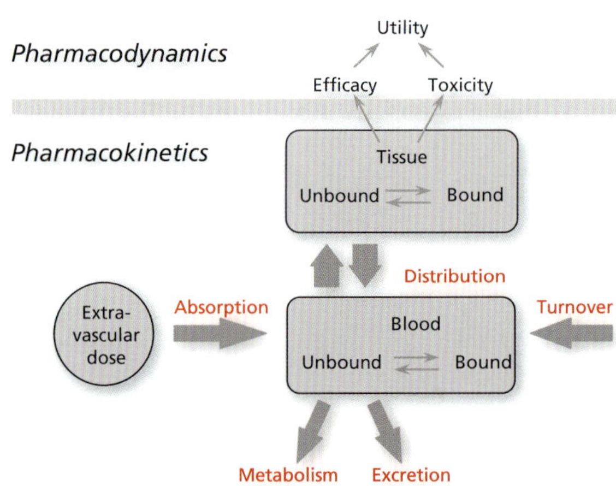

Figure 1.1
Schematic representation of the interrelationship between pharmacokinetics and pharmacodynamics. The boxes symbolize the body as being made up of a blood and a tissue compartment. Pharmacokinetics, comprising absorption, distribution, and elimination (excretion plus metabolism), is affected by turnover and binding. Excretion, metabolism, binding, and turnover may occur anywhere in the body. Turnover corresponds to the natural synthesis or secretion of endogenous compounds. Pharmacodynamics, which quantifies the efficacy and toxicity of a drug and therefore its utility, is dependent on the pharmacokinetics of the drug.

As already implied, it is commonly assumed that a pharmacological response is directly related to the administered dose of a drug. While this assumption is sometimes valid, it cannot be taken for granted. For example, many factors including bioavailability, active metabolites, and changes in plasma protein binding may confound the response. In fact, it is uncommon for all of the administered drug (*i.e.*, the dose) to be responsible for an observed pharmacological effect (response). In some cases, only a fraction of the dose reaches the plasma or blood, with the rest not being absorbed, or degraded in the gastro-intestinal tract or during the first-pass through the liver. Such degradation is a consequence of anatomy – drugs taken orally pass through the gut wall into the blood, which passes through the liver before reaching the rest of the body *via* the general circulation. Because drug metabolism occurs in both the wall of the gut and in the liver, drugs can be degraded in these sites before reaching the general circulation. There may also be time differences between drug exposure at the site of action (drug concentration at the target site) and the response due to delays introduced by biochemical and physiological processes, confounding the interpretation of the onset and the cessation of the response. It is because of complexities like these that integrating the analysis of pharmacokinetic and pharmacodynamic data is so beneficial. This is the basic tenet of quantitative pharmacology or PK/PD analysis.

Whenever possible, PK/PD analysis should be based on three types of *in vitro* or *in vivo* relationships: *concentration-time, response-time* and *concentration-response*. Figure 1.2 illustrates schematically these relationships three-dimensionally.

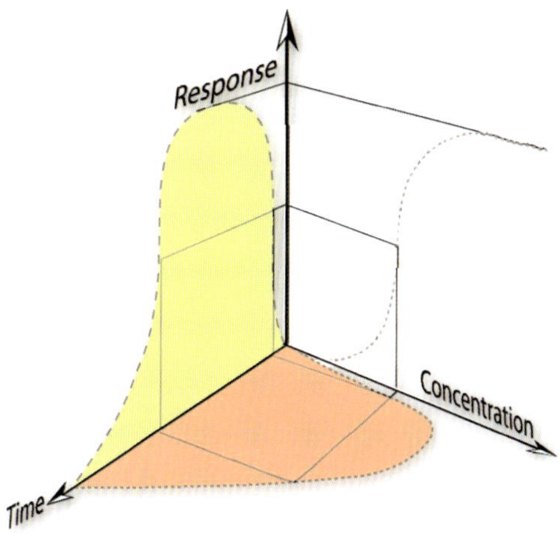

Figure 1.2
Schematic 3-D illustration of quantitative pharmacology. *In vitro* concentration-response (upper right curve) and *in vivo* plasma concentration-time (lower graph) data are combined to establish a projected *in vivo* response-time profile (upper left curve). Deviations from the projected relationship are then examined in terms of distribution, turnover–driven kinetics, receptor on/off delays, functional adaptation, synergy, and so on.

By considering the relationships between the pharmacodynamics (i.e., time-course of the biological effects of a drug, the relationship of these effects to drug exposure, and the mechanisms of drug action) on the one hand, and pharmacokinetics on the other (i.e., drug absorption, metabolism, distribution, excretion and turnover), it is possible to build mathematical models that predict, for example, the relationship between drug concentration in blood and the response in man. Relationships explored in this way provide clinically useful information about baseline and maximal drug effects (sensitivity aspects of the response), and the change in the observed effect with time (temporal aspects of the response) (see Figure 1.3). In other words, the primary objective of PK/PD modelling is to identify the key *in vivo* properties of a drug, thus expediting the prediction of the magnitude and time-course of drug response under physiological and pathological conditions in animals and, ultimately, man.

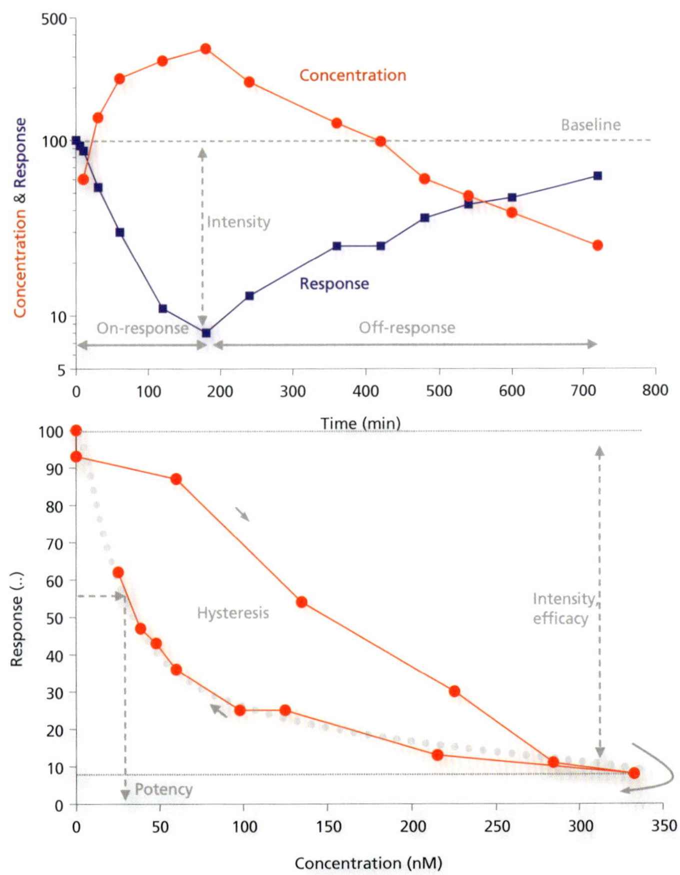

Figure 1.3
Upper: The concentration-time course (red) and response-time course (blue) after a constant rate intravenous infusion of drug to an animal. The example illustrates the pharmacodynamic characteristics during 'on' and 'off' response phases after administration.
Lower: The concentration-response relationship plotted in *time order*. This relationship shows a typical clockwise hysteresis. The turnover of response is faster than the turnover of drug in plasma which still results in a disequilibrium between the two. A good estimate of potency is obtained from the terminal portion of this loop. Potency is estimated to about 30 nM. An almost 100 % loss of response is indicated by the more than 90 % intensity. Pharmacodynamics include efficacy and potency (sensitivity aspects) and time to equilibrium (temporal aspects and hysteresis). The drug kinetics (exposure levels) are considered to 'drive' the dynamics. The dotted grey line represents the equilibrium concentration-response relationship.

1.3.2 General challenges for PK/PD in drug discovery and development

Because the pharmacological response generated by a drug is highly dependent on the actual system and conditions used to study its action, knowledge about a drug's potency and efficacy at a given concentration *in vitro*, or dose *in vivo*, is insufficient to obtain a complete understanding of its pharmacodynamic profile. During drug development, various *in vitro* and *in vivo* studies are also usually performed in order to establish the pharmacokinetic properties of the drug. Based on these, attempts are made to deduce the relationship between drug concentration and effect. However, it is not always appreciated that there is a need to couple the drug concentration in relevant compartments *in vivo* to the biological response in order to obtain a complete and meaningful kinetic-dynamic understanding. For example, an investigator may overlook drug exposure, being satisfied with dose-response data and, sometimes, with inappropriate routes of administration. However, such data do not establish the complexities of a drug's pharmacodynamic properties, which also requires knowledge about the potency and efficacy of the drug in complex integrative cellular systems, combined with evidence that *in vitro* activity translates to relevant effects *in vivo* (*e.g.*, blood pressure control). A crucial challenge, therefore, is to determine drug exposure over time in preclinical animal models of the target effect (*i.e.,* to show if the drug acts on the desired/targeted receptor population) and in disease models (*i.e.*, to show if the drug influences some manifestation of the disease). Other challenges include clarifying disparities in a drug effect due to genetic and other differences caused by species, strain, and subject variation.

The ideal pharmacodynamic measure should be sensitive, graded, meaningful, objective, and reproducible. In order to characterize the pharmacodynamics of a drug, the (baseline or disease model) control level should be included, as well as the drug response at 2 to 4 dose levels (of which one level should produce the maximum response), and the relationship between exposure and time should be adapted so as to capture the onset, intensity, and duration of the response. Ideally, the model should also incorporate the same mechanism of action intended for the drug therapy, and it should elucidate pivotal intermediate steps.

1.3.3 General terminology

There are two main types of *response*. The first, said to be *graded* or *continuous* response, may be exemplified by physiological variables such as blood pressure, heart rate, and acetylcholinesterase activity. In such responses, any change can be measured as a continuous function of dose, concentration, or time. (It should be kept in mind, however, that physiological reflexes and compensation mechanisms that sometimes confound the response read-out may need to be accounted for; *e.g.*, cardiovascular "buffering" systems). The second type of response is *quantal* or *dichotomous* response, with there being either a response or no response (*i.e.*, the concept of all-or-none), but nothing in between. Epileptic seizures, cancer, and death exemplify this group. Analysis of continuous and dichotomous responses requires different approaches.

Because it is through receptors that drugs mediate a response, receptors play a crucial role in pharmacodynamics and, therefore, quantitative pharmacology. Not surprisingly, a specific terminology around receptors has built up over the years.

In a broad sense, a *receptor* is a molecular constituent of an organism with which an endogenous or exogenous chemical/drug (sometimes referred to as a *ligand*) interacts and by so doing affects intrinsic biochemical/physiological functioning of the receptor.

Affinity and *efficacy* traditionally refer to drug action characterized in terms of ability to bind to a receptor and to alter a response *via* this interaction, respectively.

- *Affinity* (or K_d) for receptors is an independent description of the attraction between a drug and its biological target, and is thus governed by the chemical forces that govern the association between drug and receptor or receptor state.
- *Efficacy* is a measure of the degree of response resulting from the drug-receptor interaction.

The following is a simplified representation of the properties of ligand-receptor interaction. As illustrated in Figure 1.4, receptors may exist in active and inactive conformational states (R_a and R_i, respectively), and a ligand that interacts with a receptor may function as an *agonist* (conventional or *inverse*, *full* or *partial*) or an *antagonist.* Which of the aforementioned properties are displayed by the ligand is ultimately determined by its *relative affinity* for these conformational states in a given biological system.

A (conventional) *agonist* has higher affinity for the active than the inactive conformation of the receptor and will increase the proportion of receptors in the active state. Conversely, *inverse agonists* have higher affinity for the inactive conformation and drive the active-inactive receptor equilibrium towards the inactive state. In a given system whereas a *full* agonist is able to elicit the maximum possible response, a *partial* agonist will never do so, even if it occupies all of the available receptors (this applies to conventional as well as inverse agonism). An *antagonist* has equal affinity for receptor states and reduces the action of agonists; on its own, however, an antagonist is not capable of producing a response.

If in the absence of ligand the equilibrium of receptors is directed strongly but not fully towards the inactive *vs.* the active state, it may be difficult to differentiate between inverse agonism and ordinary competitive antagonism. The term *"neutral (*or sometimes *"silent") antagonist"* has therefore been introduced, to describe antagonist ligands that are able to block conventional and inverse agonists, but will not *per se* alter the basic receptor equilibrium across active and inactive states (*i.e.*, in fact, really a true conventional *antagonist*).

In some tissues, due to very efficient receptor stimulus-response coupling (transduction) mechanisms, high efficacy agonists can produce a maximal effect even if only a small fraction of the (active state) receptors is occupied. In these situations, the tissue is said to possess *spare receptors* or a *receptor reserve* for a given level of response to the agonist.

Figure 1.4
Schematic illustration of ligand-receptor interactions R_i, R_a = Receptor in inactive and active state, respectively A = ligand (agonist, antagonist, partial agonist or inverse agonist). Further explanation; see text.
Upper curve: Concentration-activity of a full agonist. Maximally possible response is reached at appropriate concentrations.
Middle curve: Concentration-activity of a partial agonist. Maximally possible response is not reached.
Horizontal line: Concentration-activity of an antagonist. There is no change in the balance between active and inactive receptors. The antagonist has no intrinsic activity. It binds to the receptor but does not activate the receptor.
Lower curve: Concentration-activity of an inverse agonist. The equilibrium is shifted towards inactive receptors causing a reduced activity.

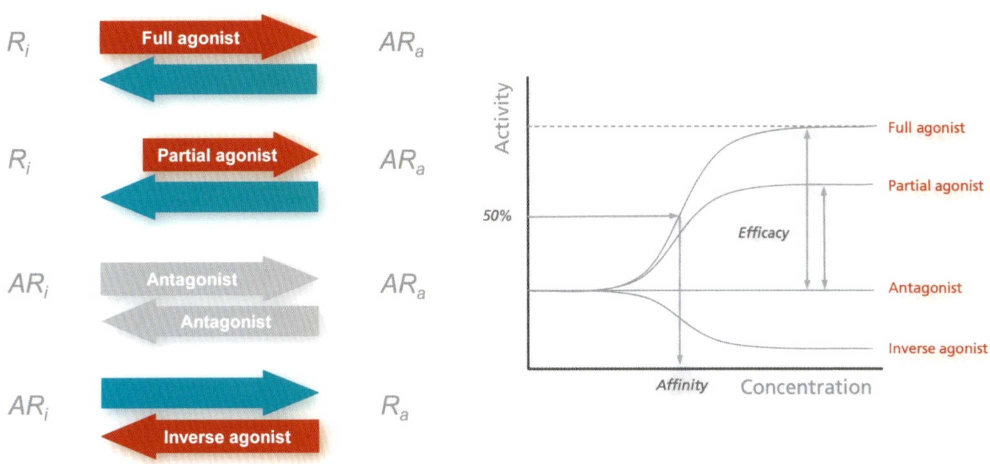

For a further, thorough discussion on basic receptor principles, theory and concepts, the interested reader is referred to reference pharmacology textbooks (*e.g.*, Goodman & Gilman, 2011, Kenakin, 2006).

Potency is an expression of the activity of a compound, in terms of either the concentration or amount (dose) needed to produce a defined level of effect. More accepted ways of expressing potency include EC_{50} which is the (preferably molar) concentration of an agonist that produces 50 % of the maximal possible effect, and IC_{50}, the (preferably molar) concentration of an antagonist that reduces a specified response to 50 % of its former value.

Figure 1.5 shows the above concepts related to receptor terminology and activation schematically.

Desensitization, tolerance, and *tachyphylaxis* refer to the spontaneous decline in a response during continuous application of an agonist, or during repeated applications of doses. It is recommended that the term desensitization be used when the reduced response is considered to involve the receptor itself, or to be a direct consequence of receptor activation.

Equilibrium between the pharmacological response and measured plasma concentration is sometimes rapid or instantaneous. This is observed as maximum response peaks at the maximum plasma concentration. A concentration-response relationship as shown in Figure 1.6 (left) can be used to predict the response-time course directly from the concentration-time course. This will be further discussed in Chapter 4. In other situations the plasma concentration-time course and response-time courses are shifted in time as shown in Figure 1.6 (right). The latter scenario, in which the drug concentration in plasma changes more rapidly than the response time-course, is probably the more common. Delayed concentration- and response-time courses will be discussed in Chapter 5.

Figure 1.5
Left: Schematic presentation of baseline, maximum response, efficacy (E_{max}) and potency (EC_{50}). **Right:** Schematic presentation of baseline, maximum response, efficacy (E_{max}) and potency (EC_{50}) of full- and partial agonists, antagonist and inverse agonist.

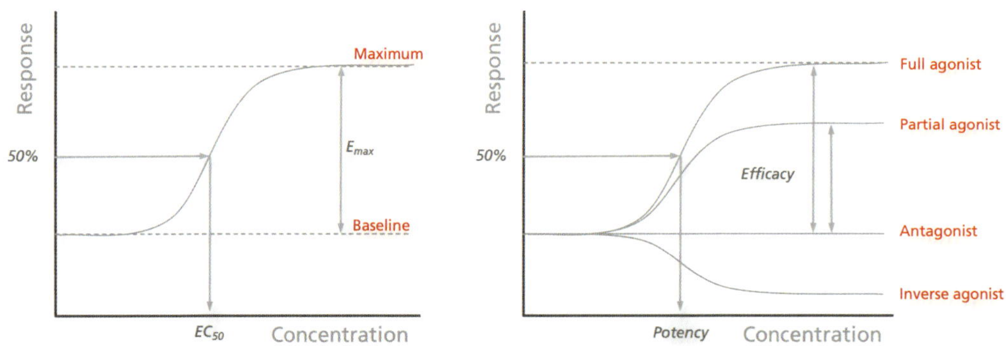

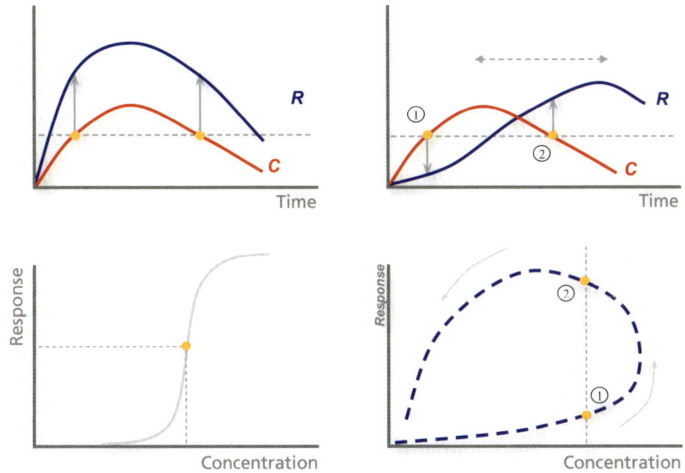

Figure 1.6
Left column: A rapid concentration-response diagram where the concentration-time and response-time courses peak at the same time (upper left) and the concentration-response relationship (lower left) thus displays no hysteresis when concentration and response data are plotted in time order. **Right column**: A situation displaying temporal differences in the respective drug concentration- and response-time courses. In this case, plotting concentration and response data in time order describes a clearcut hysteresis curve shape.

Inevitably, many basic terms will arise repeatedly during any discussion of quantitative pharmacology. All describe some basic aspect of pharmacodynamics and/or pharmacokinetics and, as such, the concepts they embrace are inherent to understanding the pharmacology of a drug. Of a more vague and sometimes individually flavoured nature, however, is the concept of target engagement, which plays a crucial role in drug development. This is introduced in the next section.

1.4 Drug discovery and development

1.4.1 The importance of quantifying target engagement

Target engagement is generally conceived of as the degree of exposure of the target area to the drug, its binding to the target, and the effect on subsequent target pharmacodynamics. It can be studied in many different ways. When studied by means of receptor occupancy (fraction of receptor population occupied by ligand), the reaction to the drug is quantified in terms of the drug's exposure-time profile (*i.e.*, its pharmacokinetics), the ligand-receptor interaction (*i.e.*, its pharmacodynamics), or the turnover of the physiological response *per se*. By *turnover* is meant the time-course of the various processes involved in the production or build-up of a pharmacological response. Such processes may be target activation, the physiological response of target activation, a pathophysiological reaction, or the clinical outcome. This time-course is governed by both production (synthesis) and loss (catabolism) processes. Take, for example, the irreversible action of the sulphenamide metabolite of omeprazole on proton pumps in the stomach. These pumps, which are made up of an enzyme known as H^+/K^+-ATPase because of its role in pumping hydrogen ions into the stomach in exchange for potassium ions, regulate stomach acidity. Even in the absence of omeprazole with its short plasma half-life of 45 minutes, there is continuous turnover of the pumps, due to their *de novo* synthesis and breakdown. When production of the pumps balances their loss, there is no net change in their number, and the system is said to be at equilibrium or baseline. If, on the other hand, pump production exceeds pump loss, more accumulate and stomach acidity increases (*i.e.*, its pH decreases); if prolonged, this can lead to gastric ulcers. When omeprazole is ingested, it is not omeprazole itself which exerts a pharmacological effect, but its sulphenamide metabolite. This binds irreversibly to the H^+/K^+-ATPase (which functions as a receptor), rendering it inactive by preventing it from exchanging hydrogen for potassium ions. Thus, the number of available active proton pumps is reduced. The higher the dose of omeprazole, the faster the pumps are inactivated and lost and, as a result, the less acid is produced. In other words, exposure to the pharmacologically active metabolite and the consequent change in acid production are dependent on the dose of omeprazole. Regardless of how it is measured (*i.e.*, as a decrease in acid secretion, an increase in stomach pH, or a decrease in the number of available proton pumps), target engagement in this example is due to the exposure time-course of the omeprazole sulphenamide metabolite, the affinity between this metabolite and the proton pumps, and the natural turnover rate and half-life of the H^+/K^+-ATPase. Note that whereas half-life provides a measure of how long a drug remains in plasma (45 min), in this example the natural lifespan of the proton pumps (*i.e.*, their half-life of 15–20 h) also affects target engagement.

Figure 1.7 summarizes the four principal factors that influence *in vivo* target engagement, namely dose, potency (measured in terms of IC_{50}), clearance of the compound, and turnover rate of the response (referred to as the fractional turnover rate or k_{out}).

During the early stages of developing a new drug it is important to observe target engagement, because this demonstrates that the test compound exerts an effect at the target site. One way of doing this is to test the compound at two or more doses and/or by using two or more routes of administration. Ideally the pharmacological response is measured repeatedly over time, although often measurements are only made at a single time point in a large number of animals. However, single time points provide a less robust design because of the number of factors influencing target engagement. For example, as shown in Figure 1.7, dose, potency and clearance all have an impact on the intensity and duration of the response in this example, but no effect on its onset of action; on the other hand, response turnover influences the onset and intensity of the response, and duration. Apart from complexities such as these, some factors may vary over time and between studies (*e.g.*, turnover, clearance, and potency), and according to disease and species (*e.g.*, turnover rate and turnover time). In short, drug parameters differ across compounds, disease states, species, strains, and studies. This highlights the importance of understanding limitations of the single time-point approach, and advantages of the more robust time-series approach which better reveals onset, intensity, and duration of a response.

Figure 1.7
The impact of dose, potency (IC_{50}), clearance, and turnover (k_{out}) on target engagement. The four panels show simulations of response-time curves where all but one measure is held constant. Changes in dose, potency, and clearance all affect the intensity and duration of the response, but not its onset, whereas changes in turnover rate affect the onset and intensity, but not the duration, of the response. The red, blue and pink curves represent, respectively, **Upper left:** doses of 1, 2 and 5 mg; **Upper right:** potencies of 0.25, 0.5 and 1.25 mg·L^{-1}; **Lower left:** clearance of 0.0228, 0.04 and 0.07; **Lower right:** fractional turnover rates of 0.06, 0.12 and 0.3 h^{-1}.

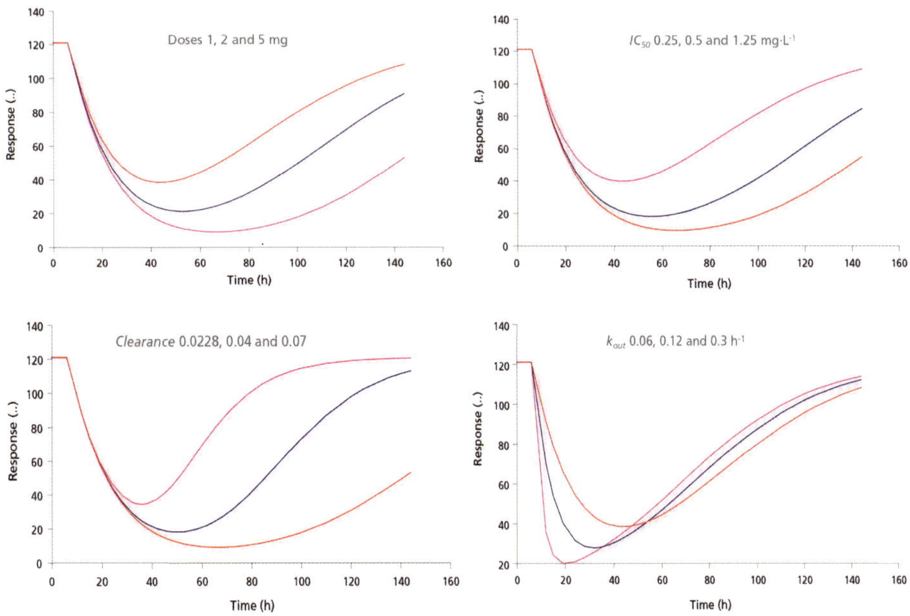

1.4.2 Major hurdles of drug discovery and development

The development of any new drug is a lengthy and expensive process. Not only is there the primary requisite to identify a medical need, but a suitable compound to satisfy this need must be identified and/or designed, and it must be formulated in such a way that it can be delivered optimally, so as to maximize target engagement and minimize undesired effects. The compound must be quantified in chemical, pharmaceutical, pharmacokinetic, pharmacological, toxicological, and clinical terms, and extensive testing is required to establish that it is efficacious and safe. In addition, the entire process must be fully documented in order to satisfy health authorities about all aspects of the testing process, and about the chemical, pharmaceutical, non-clinical and clinical properties of the final product. To achieve all of these goals, all developmental work must be carried out as efficiently, expeditiously, and cost-effectively as possible. Although admittedly difficult to assess, it has been estimated that the average cost of producing a new drug and taking it to market is around 1 billion US$ or more (Adams & Bentner, 2010), a figure that may, in part depending on variation in expenditure by therapeutic category, prove to be even higher. Drug development is a time-consuming and painstaking process with high risks of attrition. Clearly, careful and thoughtful co-ordination, planning, and study design both within and between different disciplines is thus required to avoid unnecessary duplication, time- and resource-wasting.

Pharmacokinetics and pharmacodynamics both play a huge role in drug development, right from the time a new drug target idea is conceived. For example, data obtained during preliminary screening studies, be they *in vitro* or *in vivo*, can be of major benefit to medicinal chemists as they modify and improve aspects of drug design. Likewise, experimental data influence pharmaceutical development, and subsequent preclinical and clinical studies. All of these activities have one end in mind – the successful, timely, and efficient development of a drug with good efficacy and low toxicity. Of course, this requires that the pharmacokinetics and pharmacodynamics of the compound are fully appreciated prior to its first use in man, and that data obtained from non-human species, from *in vitro* or *in vivo* studies, are accurately predictive of outcomes in man. It is in enhancing the likelihood of achieving these goals in all phases of drug development that quantitative pharmacology comes to the fore.

1.4.3 Biomarkers as navigational tools in pharmacology

To maximize pharmacological knowledge, be it for research purposes or in the context of drug development, a successful pharmacokinetic-pharmacodynamic strategy must be developed. In the context of drug development, this necessitates navigating a way through the complexities of preclinical data so as to successfully predict human pharmacology and clinical outcomes. This is achieved using *biomarkers*, a term encompassing a multitude of measures (genetic, molecular, pharmacokinetic, pharmacodynamic, physiological, clinical, and more) each of which is used for a specific purpose, and all of which are specific in terms of where they add value. For example, biomarkers include phenotypic and genotypic characteristics, plasma concentrations, measures of target occupancy and activation, physiological responses, and disease processes and outcomes. It appears reasonable to predict that translational biomarkers will become the mainstay for future tailor-made approaches to the treatment of human medical conditions (aka Personal Health Care, PHC). As indicated in Figure 1.8, there are three major considerations in any strategy that employs biomarkers:

- Firstly, that the proposed linkage provides a distinct and navigable route for testing in non-human species (testing way).
- Secondly; the outcome of measured biomarkers needs to be quantifiable (*e.g.*, clearance *Cl*, potency EC_{50}).
- Thirdly, the translation of properties determined and extracted in non-human species (*e.g.*, dose *D*, clearance *Cl*, potency EC_{50}) into predicted human properties must be justifiable.

Schemes like the one outlined in Fig. 1.8 further emphasizes the importance of utilizing PK/PD data to its fullest extent to provide the most advantageous basis for prospective therapeutics. In its most successful shape this approach can be expected to speed up drug development, reduce attrition, and increase treatment specificity.

Figure 1.8
Schematic representation of the concept of biomarkers linking genotype/phenotype (Type 0) to some clinical measure (Type 5 and 6). Type 0 biomarkers are geno/phenotypically and/or potentially disease-elicited deviations from the norm, ultimately responsible for all drug actions in that they influence disposition (Type 1, absorption, distribution, metabolism, excretion, and turnover), target occupancy (Type 2, affinity and extent), target activation and therefore function (Type 3, intrinsic efficacy and receptor expression), physiological response (Type 4, measured in the integral biological system), and disease processes (Type 5, pathophysiological measures) and outcomes (Type 6, clinical response). Model adapted from Jusko and Ebling (1995), Peck and Wechsler (2002) and Danhof et al (2005). This is exemplified by an obesity project using diet induced obese (DIO) mice as the non-human Type 0 biomarker. The blue bubbles are the testing-way in non-human species. The outcome variables ranges from clearance and exposure (in Type 1) to effective concentrations and half-life of fat turnover (in Type 6). The vertical grey arrows denote the linkage between non-human and human parameters measured (translational aspects).

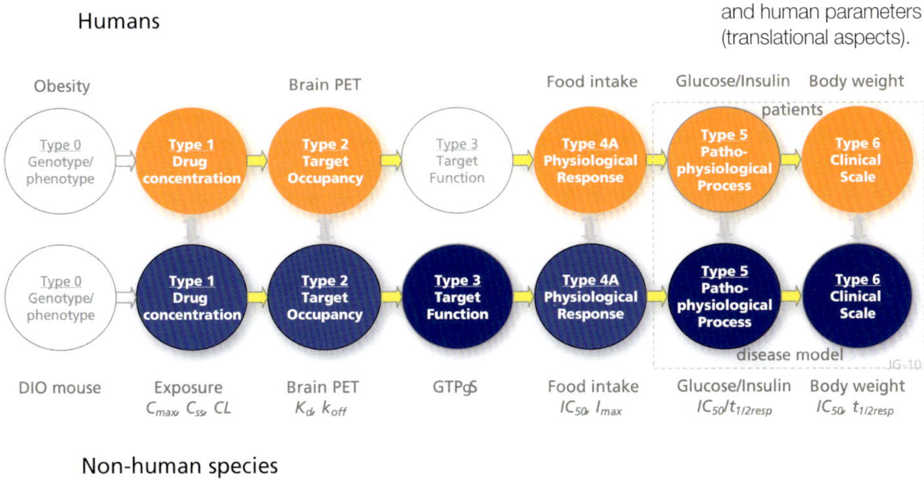

If a strategic biomarker drug development program is implemented, the outcomes will provide evidence about the proposed mechanism of action of the compound (from target occupancy and target activation data), the proposed physiological consequences of target activation, and the clinical consequences of drug exposure. Some examples of biomarkers and key design requirements of biomarker studies are shown in Table 1.1.

Table 1.1
Classification and features of different types of biomarkers.

Type	Measure	Examples of biomarkers	Outcome variables	Key design requirements
0	Genotype, phenotype	Expression of enzyme, transporter, receptor, fast- and slow metabolizer		
1	Concentration (parent drug, metabolite, *etc.*)	Concentration in plasma, tissue, *etc.*	C_{max}/C_{min}, AUC, Cl, Vss, $t_{1/2}$, F	Exposure-time data at two doses
2	Target occupancy	Degree of occupancy, affinity; PET	K_d, k_{off}	Time series of target binding at two doses
3	Target activation	Quantitative EEG parameters of $GABA_A$ receptor activation *in vivo*	K_d, k_{off}, RT	Time series for target activation at two doses
4	Physiological measure	Variables that characterize test compound effects *in vivo* (*e.g.*, blood pressure), food intake	E_{max}, EC_{50}, k_{in}, k_{out}, $t_{1/2kout}$	Response-time data at two doses after repeated dosing
5	Disease marker	Measures of inflammatory reactions, neurofibrillary tangles, glucose, insulin, leptin	E_{max}, EC_{50}, k_{in}, k_{out}, $t_{1/2kout}$	Mimic **symptoms** (pain, anxiety, cognitive impairment), **key features** (neurodegeneration, uncontrolled cell division, formation of atherosclerotic plaques), **entire process** (model of symptoms, pathophysiology)
6	Clinical scale	Progression of disease (*e.g.*, Parkinson's, Alzheimer's), body fat		Measure changes over time after repeated dosing, at two doses

1.4.4 The general consequences of quantitative pharmacology

Over the last two decades, a focus on *in vitro* high throughput assays and a "process-led" style of working changed how combined pharmacokinetic and pharmacodynamic characteristics were assessed. Very early in drug development it may be found that a new compound is efficacious *in vivo*, even though some of its *in vitro* or *in vivo* PK properties are far from optimal. For example, it may have low *in vitro* metabolic stability (high intrinsic clearance Cl_{int}), low bioavailability F, and/or a short half-life $t_{1/2}$, or perhaps high *in vitro* intrinsic clearance Cl_{int} coupled with low *in vivo* clearance Cl. If the true significance of such observations is to be understood so that compounds can be improved, the consistency between the *in vitro* and *in vivo* pharmacodynamics, between the *in vitro* and *in vivo* pharmacokinetics, and between the pharmacodynamics and pharmacokinetics must all be considered. In other words, the ultimate selection of a compound for further development should be based on the *combined value* of its pharmacodynamic and pharmacokinetic properties, as it is rare that the pharmacokinetics alone are adequate or sufficient when predicting *in vivo* pharmacology (Fig. 1.9). The importance of such a combined approach is highlighted by drugs where a clear-cut and highly desirable pharmacological effect was given precedence over problems associated with clearance and/or bioavailability (*e.g.*, felodipine, omeprazole, quetiapine).

Figure 1.9
The shared responsibility approach. Bioscience (pharmacology), DMPK and Medicinal Chemistry functions'share' the responsibility for ranking of compounds during the life-time of a discovery project. A common platform could be human dose predictions. The human dose is a consequence of a compound's pharmacological (potency, efficacy, effective concentration range) and DMPK (clearance, fraction absorbed) and system (half-life of response) properties. *CL* and ***Ce*** denote clearance and effective concentrations, and Cl_{int}, K_d and EC_{50} correspond to *in vitro* intrinsic clearance, *in vitro* dissociation constant and *in vivo* potency, respectively. Affinity, efficacy and system represent *in vitro*/ *in vivo* binding and capacity, and physiological properties, respectively.

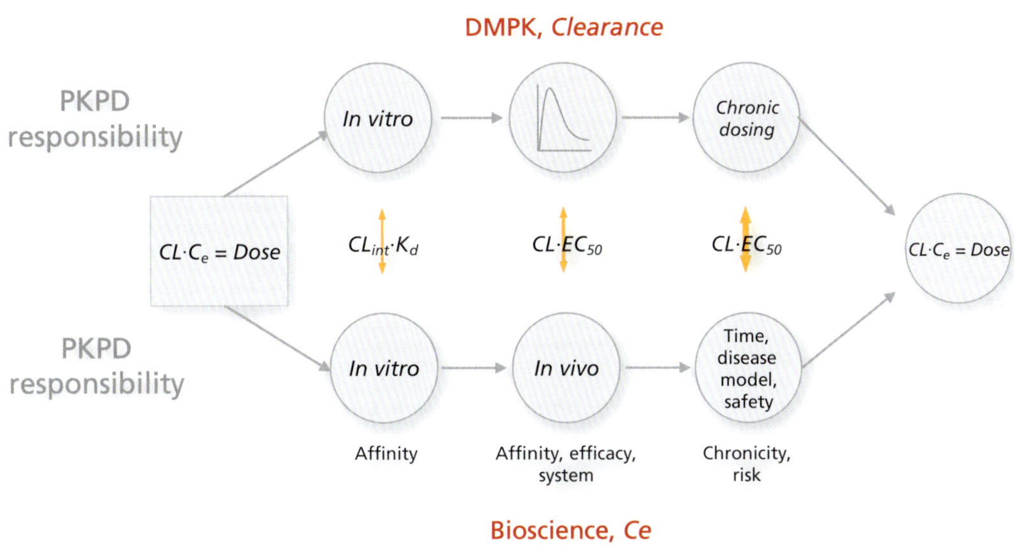

In order to help characterize the pharmacokinetics and pharmacodynamics of a compound during discovery and early development, an excellent option is to establish solid evidence for target engagement in terms of target occupancy and target activity. Because target engagement can be quantified using the different approaches of clarifying pharmacokinetic exposure-time relationships, pharmacodynamic receptor-ligand interactions, and changes in physiological systems (see Section 1.4.1), these provide a powerful tool in this regard. Clarifying the compound properties in *in vitro* binding (protein, receptor, enzyme, cell) and appropriate but fast *in vivo* pharmacodynamic screens is invaluable, as is obtaining early evidence of the effect of the compound in an appropriate *in vivo* disease model, even though disease models are frequently more time-consuming and costly. Once linkage to the disease model has been established and validated for key compounds, it is possible to focus further screening on the faster methods (*in vitro* binding, rapid *in vivo* screens). Ultimately, physiological systems will need to be clarified in terms of normal or baseline properties, and disease-affected or disease-model properties. Hence, it is necessary to investigate properties such as baseline behaviour, impact of endogenous ligands, variability, and turnover characteristics (*e.g.*, half-life $t_{1/2}$ of response), and to look at the effect of such factors on the disease, prior to testing novel compounds *in vivo*.

Failure to take into account all the known pharmacokinetic and pharmacodynamic features of a drug may not only mean that studies are poorly conceived and/or designed, but may also lead to invalid conclusions and errors in estimating the efficacy and/or risk of a compound. For example, differences in protein binding between species may blur information about efficacious concentrations, safety margins, or the shape of the concentration-response curve, and temporal differences between drug concentrations in plasma and the drug's effects may confound interpretation of the relationship between drug concentration and a pharmacological effect. Furthermore, factors such as drug-induced sensitization or tolerance may alter the magnitude of a pharmacological effect, while unidentified active and interactive metabolites may influence outcomes, as may unquantified target-mediated drug disposition or turnover of endogenous ligands. To avoid such pitfalls, assays, experiments and models at all stages of development must be appropriate or the consequences may prove disastrous, and the species and models employed must have predictive value for the patient population. Finally, it must not be forgotten that drugs may have excellent pharmacodynamics (potent, efficacious, slow turnover of effect) but sub-optimal pharmacokinetic characteristics (high clearance, low bioavailability). Thus, unless early *in vivo* studies clarify the consequences of target engagement as determined by both pharmacodynamics and pharmacokinetics, the basis for solid and rational decision-making will be substandard.

To give an example, compound *X* was used in a pivotal chronic *in vivo* study over a couple of months, to establish its disease-curing effect in animals. The outcome was exceptional, with a 45 and 90 % disease curing effect being seen at approximately 2 and 20 nM plasma concentrations at steady state (Fig. 1.10, filled red circles). Data from *in vitro* screening for binding (green triangles) and gene expression (blue squares) were collated and compared, and the relationship between the *in vivo* outcome and the *in vitro* characteristics determined. It was found that *in vitro* gene expression and receptor binding both occurred at a similar exposure to that required for a positive outcome in the *in vivo* disease model – in other words, these data showed good concurrence. When compared with the effects of compound *X* on a safety marker (in this case, liver triglycerides [purple triangles]), exposure at a level corresponding to about 45 % improvement in the disease model had minimal effect on triglyceride levels – *i.e.*, at this exposure compound *X* was highly selective for efficacy. In contrast, any further increase in positive outcome in the *in vivo* disease model lowered selectivity by increasing safety marker engagement.

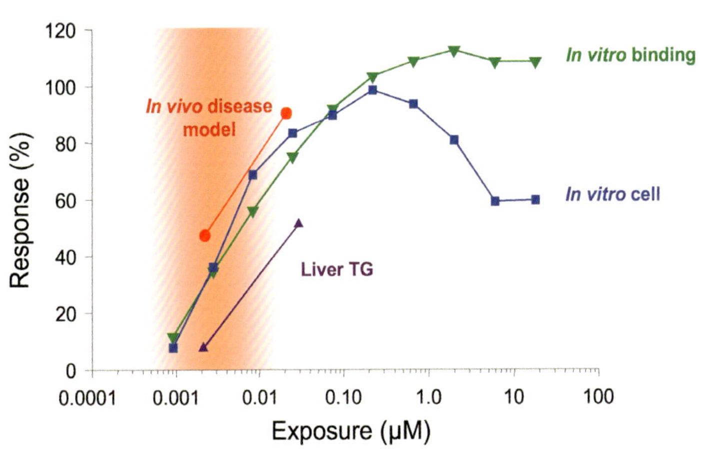

Figure 1.10
Exposure-response data for compound X in four test systems. Exposure was determined from plasma concentration, and the response in two *in vivo* (disease model, safety marker) and *in vitro* (receptor binding, gene expression) systems. Disease model improvement (red circles), safety marker (liver triglycerides - purple triangles), receptor binding (green triangles), gene expression (blue squares). The shaded area is the tentative effective plasma concentration range (1–10 nM).

This example shows that a process as simple as collating and comparing *in vitro* and *in vivo* data may reveal valuable information, in this case about both efficacy and safety. A compound's pharmacokinetic characteristics are intimately related to its *in vivo* pharmacology and safety margin (*i.e.*, the difference between the efficacious, therapeutically useful, concentration of drug, and the concentration producing unwanted/toxic side effects). However, it is surprising how frequently emphasis is placed on optimizing the pharmacokinetics *in vitro* and *in vivo*, and the target effects *in vitro*, but not on coupling these to the compound's *in vivo* properties via a holistic approach. While optimizing a property such as metabolic stability (intrinsic clearance) may improve bioavailability, it may have no effect on efficacy, and *vice versa*. In other words, too generic an approach may be detrimental, leading, for example, to the halting of a potentially successful program in early development, or to the prolongation of a program with a poorly conceived dose regimen. In short, the benefits accruing from an integrated pharmacokinetic-pharmacodynamic approach are vast, as are the disadvantages and cost of not doing so.

Clearly, a co-ordinated strategy that employs carefully designed biomarker studies is a powerful tool available to the pharmaceutical industry. By improving the processes that facilitate a thorough understanding of the relationships between a compound's target engagement, and its safety, toxicity and exposure profiles, the chances are increased of developing successful drugs in a timely and cost-effective manner. However, there are two major challenges: firstly, identifying the optimal pharmacodynamic properties of potential drugs, and secondly, clarifying the pharmacokinetic properties needed to support their pharmacodynamics. Although the current trend is to obtain most data from animal studies, health authorities world-wide want the emphasis to switch from *in vitro* to *in vivo* data collection. While this exacerbates some of the difficulties, it also provides an opportunity for a new way of working. To address these issues, we begin our discussion of the advantages to be gained from quantitative pharmacology by looking at the principles of pharmacokinetics from the point of view of pharmacodynamics.

Drug- and system-specific components interplay in portraying the overall fate and action of drugs in a living organism. Figure 1.11 shows the thread from dosing to clinical outcome which comprises drug disposition, biophase equilibration, target engagement, signal transduction and homeostatic and disease process factors. As evident from Figure 1.11, there are many levels and processes throughout the interconnected PK-PD chain of events that will influence the final response. The drug-specific element encompasses properties affecting disposition and target interaction (exposure, target affinity, occupancy, and intrinsic efficacy), whereas system-specific factors include tissue and physiological-pathophysiological processes (target density, signal transduction linking occupancy to response, disease-related changes). Drug disposition may change by disease, which further illustrates the dynamics of compound- and system-derived inter-dependence. Homeostatic adjustments to target

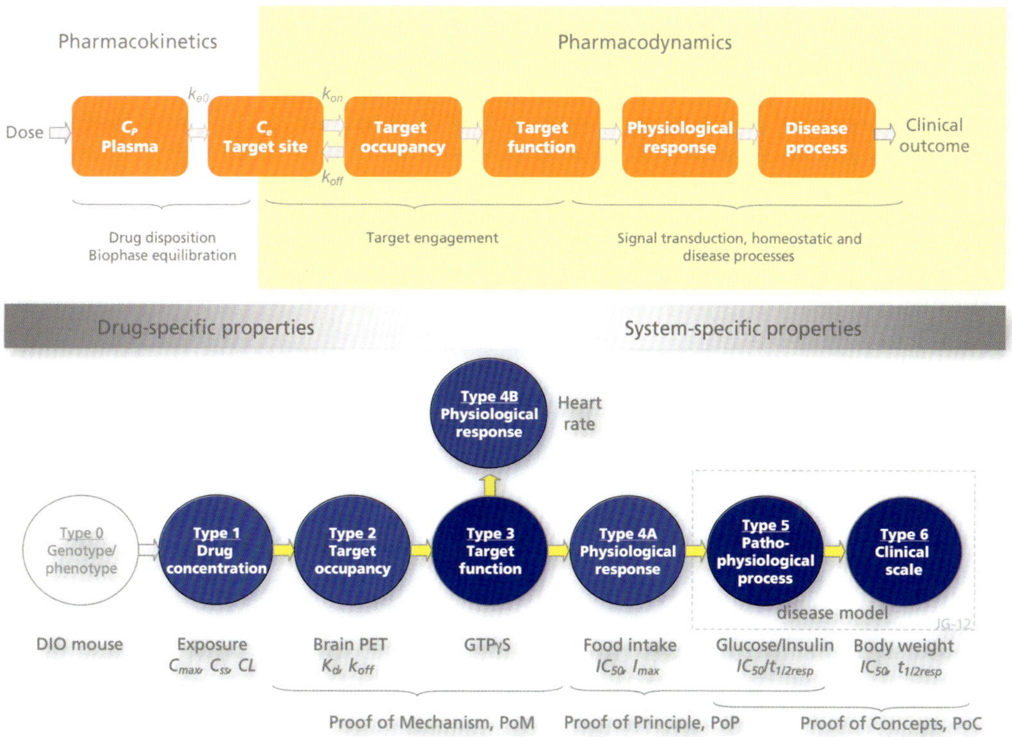

Figure 1.11
Upper part: Schematic illustration of the components of pharmacokinetic and pharmacodynamic models and how these parts are related to drug-specific and system-specific components. Generally, drug-specific components denote clearance and affinity or binding to the target. The drug plasma clearance and binding affinity of its metabolite are typical drug-specific properties. System-specific components are related to the post-target physiological and pathological events. System-specific parameters such as turnover of target and post-target factors and their corresponding biological half-lives should ideally be drug-independent. The turnover rate of proton-pumps and their corresponding half-life are system-specific properties. **Bottom part:** Schematic illustration of the biomarker classification system. The definition of the respective biomarker bubbles are found in Table 1.1. Proof of mechanism, proof of principle and proof of concepts represent different portions of the biomarker cascade.

exposure and action may also occur, such as metabolic enzyme induction and receptor tachyphylaxis.

Plasma and target site concentrations: The route and mode of administration influence drug exposure, and in turn pharmacodynamic response. Most drugs will need to enter specific tissues before exerting their therapeutic action, and penetration into the appropriate biophase is thus an important factor that needs to be factored in if (free) plasma levels of drug do not provide a straightforward correlation to drug effects. The plasma and target effect site concentrations may be viewed as being mainly governed by drug-specific properties, allowing for adequate access to the biophase.

Target occupancy and function: The relation between ligand-target binding and the induction of a biological response may deviate from linearity, presenting one level of detachment of exposure from response (see Chapter 5). Furthermore, even with a rapid equilibrium and close relationship between plasma and biophase concentrations of drug, the observed response may be significantly delayed or distorted magnitude- or duration-wise, compared to what was predicted. This obviously represents a challenge to the establishment of PK/PD relationships. For example, increased target residence time may confound exposure/response relations such that an *in vivo* response greatly outlasts what would have been predicted based on compound half-life. The extreme case of slow off-rate is represented by irreversible binding of ligand to target, and subsequent inactivation, like, *e.g.*, in the case of irreversible enzyme inhibition. The net pharmacodynamic efficacy will be a composite of the fraction of enzyme population being inhibited vs. the spare capacity in the very enzyme pool, and the actual regeneration of enzyme (determined by tissue factors, disease processes). An integrated understanding is therefore required of both PK and drug-receptor interaction, to establish a clear PK/PD association.

Physiological response and disease process: The biological response produced reflects the net effect of drug-target interaction, and is mainly determined by the actual system in which the compound acts. It is not uncommon however, to observe impaired or exaggerated response amplitudes as a result of disease processes. Sometimes more than one signalling pathway may be prevalent, the internal balance among which may be different in different tissues, and may also vary among different drugs interacting with the target receptor due to differing affinity for its conformations.

Chapter 2

Principles of Kinetics from a Pharmacodynamic Point of View

2.1 What is this chapter about?

A basic understanding of pharmacokinetic processes and of the relationships between kinetics and dynamics is fundamental to quantitative pharmacology. In order to explain how these processes and relationships can be examined mathematically, this chapter

- begins by introducing from a pharmacodynamic point of view the basic concepts, definitions, and relationships essential to pharmacokinetics, including clearance, volume of distribution, half-life, bioavailability, exposure and steady state
- develops the relationships between these concepts, where special emphasis is placed on the relationship between concentration and time, and on the area under the concentration-time curve
- discusses the significance of dosing regimens and of active drug metabolites is introduced
- introduces the basic principles underlying kinetic and dynamic models, developed and discussed in more detail in later chapters. In pharmacokinetics, the characteristics of the models that explain a the properties of a drug determine its description (*e.g.*, one-, two-, or multi-compartmental).

2.2 Essential pharmacokinetic concepts and relationships

Clearance Cl, a measure of the body's capacity to remove a drug, influences exposure (*i.e.*, the concentration of available drug at the receptors at which the drug acts) and the dose of drug required for effect. It is a virtual measure defined as the volume of plasma or blood that is totally cleared of its content of drug per unit time, and therefore has units such as mL·min^{-1}. It may also be viewed as the fraction of drug extracted across an organ multiplied by the blood flow through that organ. At some time after a drug is given, its concentration in plasma reaches a maximum, after which it eventually decreases back to its trough value. The precise pattern of a drug's concentration-time course is determined by many factors, and is thus of major interest to pharmacokineticists. However, only the principle of clearance *per se* is further elaborated on at this point, with its complexities being discussed later.

If for the sake of simplicity it is assumed that a drug requires a minimum plasma concentration to be able to exert its desired effect, it becomes clear that the lower the dose the less likely the drug is to reach this required minimum concentration (*i.e.*, the less likely it is that exposure will be adequate to produce a quantifiable effect). Furthermore, if exposure is adequate to produce an effect, the faster the drug is removed from plasma, the shorter the intensity and duration of this effect will be. These concepts are shown schematically in Figure 2.1 for a drug with simple kinetics.

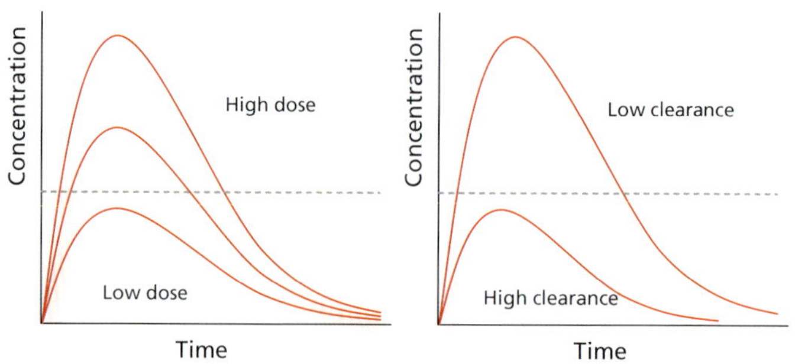

Figure 2.1
Schematic representation of the effect of changes in dose and clearance on the plasma drug concentration-time relationship. The minimum exposure required for efficacy is indicated by the dashed horizontal line. The left panel shows the relationship for three nominal doses (low, intermediate and high) of drug, and the right panel two nominal clearance values (rapid, slow). The drug concentration C reaches a maximum, before declining to zero. The lower the dose or more rapid the clearance, the shorter is the time during which the drug exerts an effect (= duration of exposure above horizontal efficacy line); if exposure is too low, there will be no detectable effect.

We talk about *systemic* exposure when drug enters the blood stream. The exposure can be expressed either as the peak plasma concentration, the complete plasma concentration-time profile or the Area Under the plasma concentration-time Curve (*AUC;* Figure 2.2). This latter area generally increases with increasing doses and one strives for estimating the *dose-AUC* relationship which may be linear or nonlinear. The *AUC* is mathematically estimated as the integral of the plasma concentration-time curve. Systemic exposure can either be presented as the total or unbound plasma concentration. For further discussions of unbound plasma concentrations see Chapter 3.

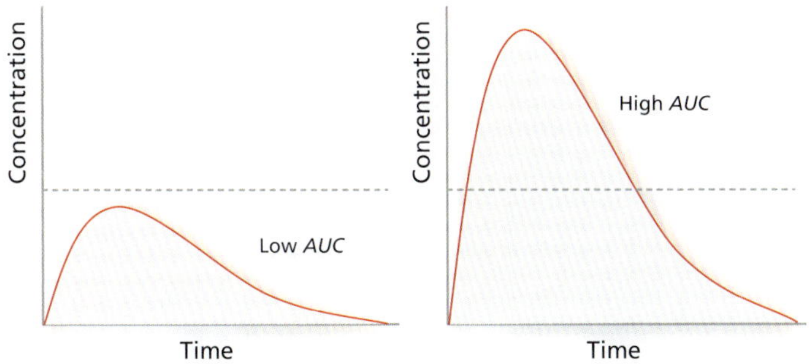

Figure 2.2
Schematic illustration of area under the curve.
Left: A low dose gives a low *AUC*, which in this case falls the systemic exposure below the target plasma concentration indicated as a dashed horizontal line.
Right: A large dose generally gives a large *AUC*, which in this case falls above the target plasma concentration for a certain duration of time.

This brings us to *volume of distribution* (*V*), a proportionality term relating plasma concentration (*C*) to the total amount of drug in the body (A_b). Because the amount of drug can also be equated with the dose of drug (*D*), volume of distribution may be expressed in two ways:

$$V = \frac{A_b}{C} \text{ and } V = \frac{D_{iv}}{C_0} \qquad (2:1 \text{ and } 2:2)$$

In Equation 2:2, D_{iv} represents the intravenous dose and C_0 the concentration at time zero. While it is commonplace to use gram-based (*e.g.*, mg, μg, etc) units for mass amount of drug, this may confound and confuse interpretation when comparing agents with significantly different molecular size and weight – including situations when a drug is metabolized *in vivo* to an entity with markedly altered molecular weight *versus* its parent. In addition, data from *in vitro* assays are most commonly expressed in molar rather than (m or μ)g units. The preferred unit for amount of drug is thus molecular mass (*e.g.*, moles rather than mg), for dose moles·kg^{-1}, and for concentration mass per unit volume (*e.g.*, moles·L^{-1}). Hence, as illustrated by Equations 2:1 and 2:2, volume of distribution has units of volume when calculated in terms of amount of drug (*e.g.*, L), or volume per unit body weight if expressed relative to body weight (L·kg^{-1}). Volume of distribution is considered to be a virtual term as it has no direct anatomical correlate. It represents a physiological component, plus the physicochemical component of tissue-to-blood partitioning. In reality, it represents the apparent space or volume into which a drug distributes after dosing, and is indicative of storage of drug in the body. It provides an important guide when assessing the tissue penetration of a drug.

Figure 2.3 shows schematically a way of visualizing volume of distribution. Two buckets of the same size are filled with 10 L of water and a small quantity of active charcoal is dispersed in the right bucket, with no significant effect on its overall volume. A 100 unit dose of drug is then added to each. After this has dissolved the concentrations are measured and found to be 10 units·L^{-1} of water in the left bucket and 1 unit·L^{-1} in the right bucket. Assume that no elimination of drug occurs from either bucket. By dividing the dose (100 units) by the measured concentration in each bucket, the volumes become 10 L for the left bucket and 100 L for the right bucket. The explanation for this apparent difference is that some of the drug became bound to the charcoal in the right bucket, which made it inaccessible to the matrix of measurement (water); as a result, the measured drug concentration in the right bucket was low and the calculated volume high. However, each bucket still contains 100 units of drug because no drug was eliminated. In other words, the volume of distribution is apparent and not real, relating the total amount of drug in the system (in this case a bucket) to the concentration at its matrix of measurement (in this case water).

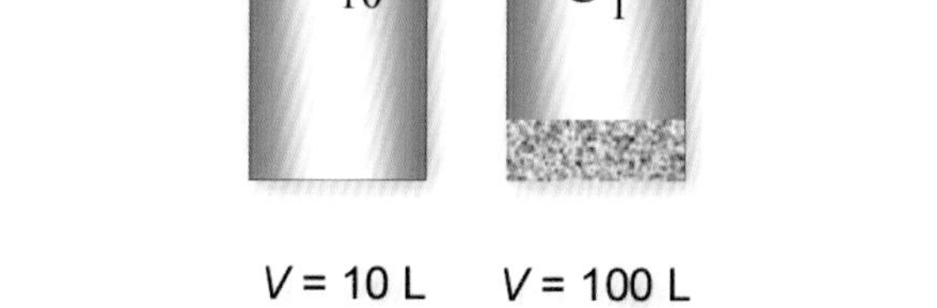

Figure 2.3
Schematic illustration of the volume of distribution, *V*. A dose of 100 units of drug is dissolved in 10 L in each of two buckets, of which the right bucket also contains active charcoal (grey area at bottom of bucket). The measured concentration *C* is found to be 10 and 1 unit·L^{-1} in the left and right buckets, respectively. Although the actual physical volume of each bucket is 10 L, the apparent bucket volumes are 10 and 100 L, respectively, because 90 % of the dose resides in the active charcoal in the right bucket.

Together, the combined effect of clearance and volume of distribution determines how long a drug stays in the body after it is administered. Because of the impracticality and inaccuracy of trying to determine the exact time at which the last molecule of an administered drug is finally removed from the body, the time used in pharmacokinetic studies is a measure of how long it takes for the drug concentration to be reduced to 50 % of a given value (*e.g.*, from a concentration of, say, 10 to 5 μmoles·L^{-1} or from 5 to 2.5 μmoles·L^{-1}). This time is called the biological *half-life* $t_{1/2}$, and it has the unit of time (min, h, *etc.*). Figure 2.4 shows schematically how it is determined.

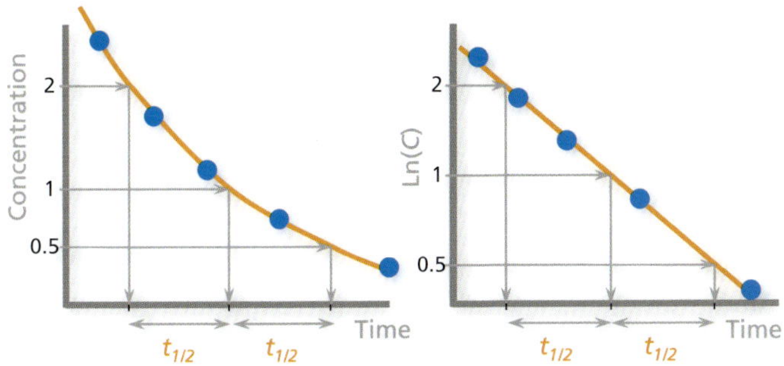

Figure 2.4
Schematic illustration of a plasma drug concentration-time profile showing how half-life $t_{1/2}$ can be determined. Note that the determination of half-life can be made from either a linear plot (left panel) or a semi-log plot (right panel). C, concentration (2 at time 1, and 1 at time 2); t, time.

The relationship between half-life $t_{1/2}$, volume V and clearance Cl is expressed mathematically as follows:

$$t_{1/2} = ln(2) \cdot \frac{V}{Cl} \qquad (2:3)$$

When drugs are given via an extravascular route (*e.g.*, oral), commonly not all of the administered dose will be absorbed, and, thus not reach the systemic circulation either. The fraction of an extravascular dose that does reach the systemic circulation provides a measure of the *bioavailability* F of the drug. Together with clearance, bioavailability is a key parameter enabling calculation of the oral dose required to achieve a specific plasma concentration. Because it represents a fractional ratio, it has no units *per se*, although it is expressed in percentage terms.

When a drug is given by constant rate infusion and if the duration of infusion is adequate, the plasma drug concentration increases to reach a plateau. The time it takes to reach this concentration depends on the half-life. In practice, it takes 3–4 half-lives to reach 90 % of the plateau (steady-state) concentration. This may also occur if drug is administered repeatedly, depending on the dosing interval. During the plateau period, referred to as *steady state*, the rate of drug input is the same as its rate of elimination. Concentration is said to be at steady state C_{ss} if the maximum concentration C_{max}, area under the plasma concentration curve *AUC*, or average concentration is the same at two (or more) consecutive dosing intervals. It is used as a measure of exposure to a drug when comparing two consecutive time periods such as dosing intervals. The concentration may still fluctuate between the two doses.

Table 2.1 summarizes the above concepts, and includes a brief description of ways in which each measure can be applied.

Table 2.1 Some key pharmacokinetic measures.

Symbol	Name: description	Units	How calculated*	Applications
Cl	**Clearance:** measure of body's capacity to remove drug	volume per unit time ($mL \cdot min^{-1}$)	$\frac{D_{iv}}{AUC_{iv}}$	To predict dose rate, exposure; to scale data from one species to another, including man; to predict human dose
V	**Volume of distribution:** proportionality constant between total amount of drug in body and plasma concentration	volume (L or $L \cdot kg^{-1}$)	$V = \frac{A_b}{C}$ and $V = \frac{D_{iv}}{C_0}$	To predict extent of distribution; to predict loading dose†
$t_{1/2}$	**Plasma half-life:** time needed for concentration to fall by 50 %	time (min or h)	$t_{1/2} = ln(2) \cdot \frac{V_{ss}}{Cl}$	Time to steady state, plasma concentration fluctuation
F	**Bioavailability:** fraction of dose that reaches systemic circulation	None; expressed as a fraction or %	$F = \frac{AUC_{po}}{AUC_{iv}} \cdot \frac{Dose_{iv}}{Dose_{po}}$	To predict oral dose needed to reach target plasma concentration
ss	**Steady state:** rate of drug input equals rate of drug elimination	Depends on parameter	$C_{ss} = F \cdot \frac{Dose\,rate_{po}}{Cl}$	At different doses, C_{ss} gives measure of exposure at given drug effect

* *A_b, amount of drug in the body; AUC, area under curve; C, concentration (C_0, at time zero; C_{ss}, at steady state); Cl, clearance; D, dose; F, bioavailability; iv, intravenous; po, oral; $t_{1/2}$, half-life; V, volume of distribution (V_{ss}, at steady state).*

† *If the initial dose of a drug is higher than subsequent maintenance doses, it is called a loading dose.*

2.3 Basic principles of compartmental modelling

The term disposition is used to encompass the processes of distribution, elimination and turnover of a drug. When a drug is given rapidly as a single intravenous injection (*i.e.*, as an intravenous bolus dose) and the concentration of drug in the plasma is determined at different times after dosing, it is possible to capture the disposition of the drug, completely avoiding any complexities of absorption as would occur, for example, after oral administration. Because rapid intravenous administration captures pure drug disposition, it is an essential tool in pharmacokinetics. For comparison, the overall kinetics of drugs that are given by extravascular dosing is given by absorption plus disposition.

Sometimes the decline in plasma concentration of drug is linearly proportional to time. More often, however, this relationship varies with time, being dependent on the plasma concentration. In fact, the rate at which the plasma drug concentration decreases is usually most rapid when the concentration is highest and wanes as the concentration declines. In other words, the concentration decreases exponentially. When the concentration-time relationship is linearly proportional, it is said to follow *zero-order* kinetics; when it is exponential it is said to follow *first-order* kinetics. Figure 2.5 illustrates the concept of zero- and first-order elimination schematically.

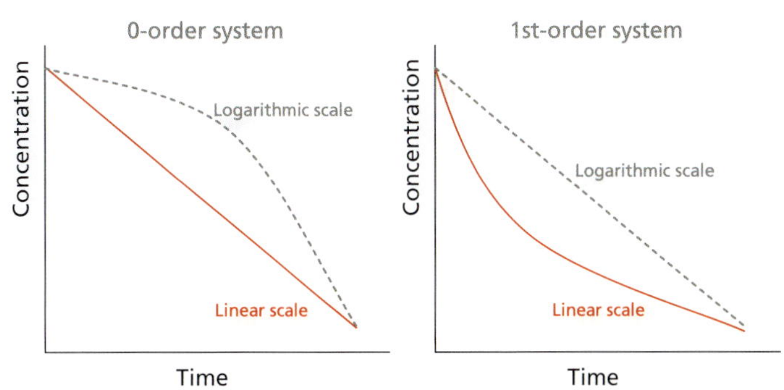

Figure 2.5
Examples of zero- and first-order elimination. Note the linear decline in plasma concentration C with time when elimination is zero-order (left panel) and the mono-exponential decline when it is first-order (right panel). Both profiles are shown as linear-linear and log-linear transformed plots.

Pharmacokinetic systems are said to be one-, two-, and multi-compartment systems. In a *one-compartment* system the entire organism is approximated to a single volume which is equated with plasma (see Figure 2.6, left). Even though the organism consists of numerous tissues and organs, it can be considered as if it were a single compartment as long as the rate at which the drug concentration changes in each tissue or organ is proportional to its rate of change in the source compartment. If this is so, and if the rate at which drug is transferred from one part of the body to another can be modelled by first-order kinetics, then the rate at which it is removed from the body as a whole can also be modelled by first-order kinetics. In fact, one-compartment models are common in pharmacokinetics because with many drugs the rate of change in drug concentration in different parts of the body is proportional. Because the number of parameters included in a one-compartment system is minimal, simple relationships can be developed between dose D, plasma concentration C, volume of distribution V, and clearance Cl. In a *two-compartment* system (see Figure 2.6, middle), the presence of an additional compartment with different kinetics adds complexities, as does a *multi-compartment* system where there are three or more compartments, each with its own kinetics (see Figure 2.6 right). All of these model types are considered in the following pages and chapters.

Figure 2.6
Schematic illustration of one- (left), two- (middle), and multi- (right) compartment models. The difference between each sort of model relates to the number of distinct phases in a semi-logarithmic plot. There is a corresponding number of compartments with different kinetic properties. Multi-compartment models can be extremely complex, with many more than the three nominal compartments shown here.

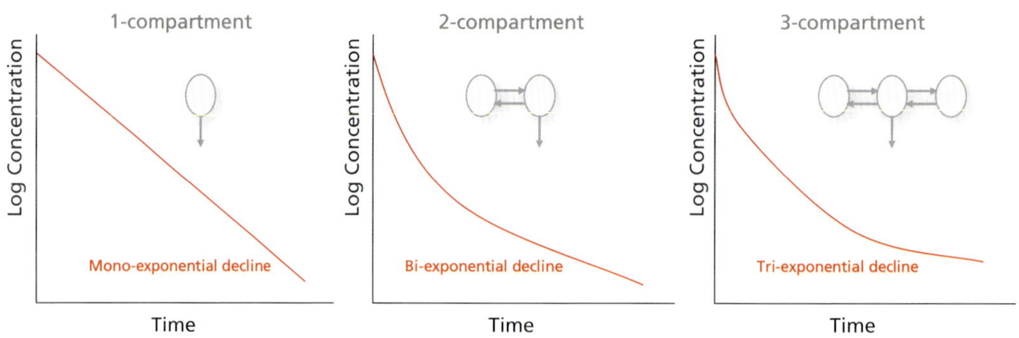

A basic understanding of drug absorption, distribution, elimination and turnover and the relationship between kinetics and dynamics, as well as the underlying mathematics, is a fundamental aspect of pharmacokinetic/pharmacodynamic PK/PD reasoning and analysis. Below, the background is given to some of the kinetic and dynamic models that are discussed in later chapters. The next section is about compartmental analysis, which is combined with various input regimens.

2.4 One-Compartment Models

2.4.1 Intravenous bolus administration

Intravenous bolus dosing, which means that all of the dose is given as a rapid injection, captures the pure disposition of a drug. Disposition stands for distribution, elimination and turnover processes. Disposition combined with absorption describes the overall kinetics after extravascular dosing. In general, the transfer rate of a drug from one compartment to another can be modeled by first-order kinetics. This means that the rate of change of drug from one specific compartment to another is proportional to its concentration within the source compartment. Figure 2.7 shows the time-course of drug in plasma of a one-compartment system where the whole organism is approximated to be a single volume. A drug that obeys one compartment disposition, exhibits a mono-exponential decline in plasma. *C* and *V* denote concentration in plasma and the volume of distribution, respectively.

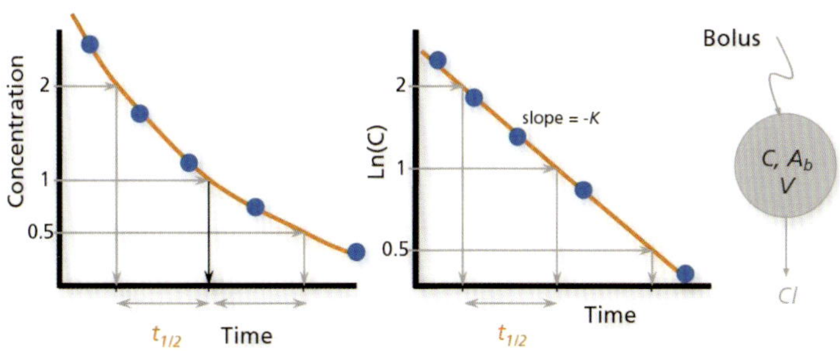

Figure 2.7
First-order kinetic data (blue solid circles) depicted on a linear scale (left plot) and on a semi-logarithmic scale (right plot). The concentration declines mono-exponentially with time in a one-compartment system. The drawing to the right schematically illustrates a one-compartment model with first-order elimination, where the concentration in plasma, amount of drug in the body, and volume of distribution are denoted *C*, A_b, and *V*, respectively, and *Cl* denotes plasma clearance.

Administering a bolus dose of a drug that behaves like the one-compartment system presented schematically in Figure 2.7, the decline in a reference volume (such as plasma), a mono-exponential concentration time course as shown in Figures 2.8 and on a semi-logarithmic scale in Figure 2.9. Note the different characteristics of the two plots.

Figure 2.8
Plasma concentration-time curves of a drug that displays one-compartment kinetics, plotted linearly. $T_{1/2}$ denotes the half-life. The behavior of Equation 2:4 is shown by the solid black line in both plots. The typical mono-exponential decline of a drug following a 10 mg bolus dose with a C_0 of 1,000 μg·L^{-1} and a half-life of about 35 minutes. The semi-logarithmic scale in Figure 2.9 reveals more clearly the number of declining phases of the concentration-time curve, and the linear scale demonstrates the time it takes to reduce the concentration by say 90 %. The exponential equation shows that the time $t_{1/2}$ it takes to lower the plasma concentration to half (50) of its initial value (100) is the half-life. The relationship between the elimination rate constant *k* and the half-life, and between half-life and the two primary pharmacokinetic parameters *Cl* and V are also shown.

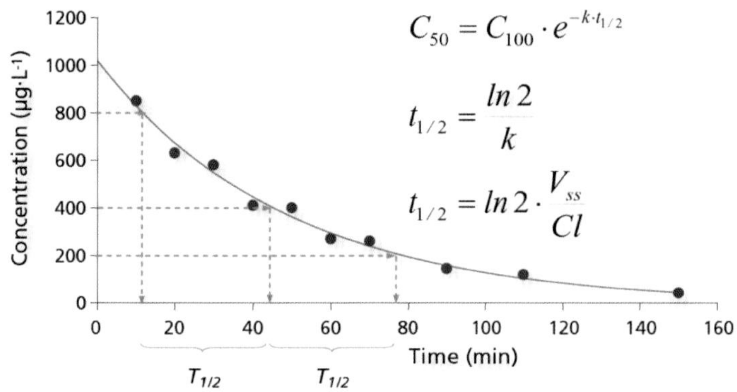

$$C_{50} = C_{100} \cdot e^{-k \cdot t_{1/2}}$$

$$t_{1/2} = \frac{ln\,2}{k}$$

$$t_{1/2} = ln\,2 \cdot \frac{V_{ss}}{Cl}$$

Figure 2.9
Plasma concentration-time curves of a drug that displays one-compartment kinetics, plotted semi-logarithmically. $T_{1/2}$ denotes the half-life. The behavior of Equation 2:4 is shown by the solid line. The typical mono-exponential decline of a drug following a 10 mg bolus dose with a C^0 of 1,000 μg·L^{-1} and a half-life of about 35 minutes. The semi-logarithmic scale reveals more clearly the number of declining phases of the concentration-time curve, and the linear scale demonstrates the time it takes to reduce the concentration by say 90 %.

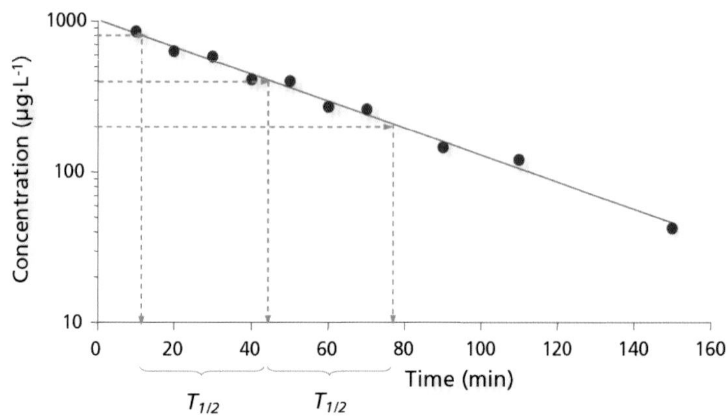

The relationship between the concentration C and time when drug is administered as a bolus dose, may be expressed mathematically for a one-compartment model with first-order kinetics as follows

$$C = C^0 \cdot e^{-K \cdot t} = \frac{D_{iv}}{V} \cdot e^{-\frac{Cl}{V} \cdot t} \quad (2:4)$$

C is the plasma concentration and K is the first-order rate constant associated with the elimination process. The solid line in Figures 2.8 and 2.9 shows the behaviour of Equation 2:4. This Equation is actually a mathematical interpretation of many first-order mono-exponential processes in the body. Applying values to the above exponential relationship to give, say,

$$C_{400} = C_{800} \cdot e^{-K \cdot t_{1/2}} \text{ or } C_{100} = C_{200} \cdot e^{-K \cdot t_{1/2}} \quad (2:5)$$

it is seen that the time it takes for the plasma concentration to decrease to half (400 or 100 μg·L^{-1}) of its previous value (respectively, 800 or 200 μg·L^{-1}, respectively) corresponds to a half-life, $t_{1/2}$, of about 35 min. In the right hand part of Equation 2:3, the clearance *Cl* is also introduced. Clearance is an imaginary term, describing the capacity for permanent elimination of a compound from plasma, and is defined as the volume of blood or plasma that is totally cleared of its content of drug per unit time. Its usefulness will be demonstrated in several examples later in this book. However, it is a consequence of a combination of perfusion, diffusion, filtration, metabolism and transport processes of the clearing organ(s), and it relates the rate of elimination *dX/dt* to plasma concentration *C*. *dX/dt* denotes the change in amount *X* per unit time, *dt*. This can be seen in the following equation:

$$\frac{dX}{dt} = -Cl \cdot C \quad (2:6)$$

The amount eliminated over time t is often expressed as a ratio *dX/dt*. The minus sign in Equation 2:6 indicates that the rate of elimination decreases as the plasma concentration C decreases over time after a single intravenous injection. Clearance is then the proportionality parameter between plasma concentration and rate of elimination. If for example 10 mg of drug is eliminated over a 2 hour period, then the elimination rate *dX/dt* is on average 5 mg·h^{-1}. If we assume that the drug is given repeatedly with a fixed dosing interval, steady-state is reached when two consecutive intervals display the same plasma concentration-time course. The areas under the two plasma concentration-time curves will also be the same.

Clearance relates the rate of elimination dX/dt to C as in Equation 2:6 and Figure 2.10. The rate of elimination and plasma concentration-time curve can be integrated over a dosing interval, yielding the total amount being eliminated and the area under the plasma concentration-time curve (AUC) within that interval. Thus, the $AUC_{0\text{-}24}$ corresponds to the amount of drug being eliminated from the body between two doses separated by 24h.

Clearance (removal)

$$Elim\ \ rate = Cl \cdot C = Dosing\ rate$$

Cl (vol/time)
C (amount/vol)
Rate (amount/time)

$$\int_0^{24} Elim\ \ rate = Cl \cdot \int_0^{24} C_{ss} = \int_0^{24} Dosing\ rate$$

$$Amount\ elim = Cl \cdot AUC_0^{24} = Dose$$

Area (conc · time)
Dose (amount)

Figure 2.10
The elimination rate is determined by clearance *Cl* and plasma concentration *C*. At steady-state, the elimination rate is equal to the dosing rate. The integral of the elimination rate during a dosing interval of e.g., 24 h, is equal to clearance multiplied by the integral (area) of the plasma concentration-time course during the same period. The total amount eliminated is equal to the total dose administered within a dosing interval at steady-state; that is, the plasma concentration remains the same before and after the dosing interval.

In other words, clearance Cl is either the proportionality term between rate and plasma concentration, or dose and area under the plasma concentration-time curve, AUC. The latter two (dose and area) are integrals of rate (amount per time) and concentration (amount per volume), respectively.

Figure 2.10 demonstrates that the clearance term can be used to relate what is being eliminated within a dosing interval at steady-state (after repeated dosing) to the area within that dosing interval. This is the same as saying that clearance is the ratio between what has been eliminated after a single dose and the area under the plasma concentration-time curve from the dosing time to infinity. The total area under the plasma concentration-time curve from time zero to infinity is proportional to the administered dose.

2.4.2 Extraction and organ clearance

Besides viewing clearance as a *Dose*-to-*AUC* ratio, one may also look at clearance from a more physiological perspective. The drug can be eliminated or cleared by the liver and the kidneys, referred to hepatic and renal clearance, respectively. Total clearance is the sum of hepatic and renal organ clearances. In this context, hepatic and renal clearances are the products of their respective *perfusion rates* (blood flow) and *extraction ratios*. The *extraction ratio* (E_H or E_R for hepatic and renal extraction ratio, respectively) is defined as the concentration difference between incoming (C_{in}) and outgoing (C_{out}) concentrations of an eliminating organ divided by C_{in}. This ratio varies between zero and 1 (unity) depending on the clearing capacity, such as the metabolic capacity or filtration rate of the clearing organs. The hepatic extraction ratio thus becomes

$$E_H = \frac{C_{in} - C_{out}}{C_{in}} \tag{2:7}$$

The extraction ratio is independent of the perfusion rate (Blood Flow; Q_H) through the eliminating organ. The hepatic organ clearance therefore becomes

$$Cl_H = Q_H \cdot E_H = Q_H \cdot \frac{C_{in} - C_{out}}{C_{in}} \tag{2:8}$$

Since hepatic organ clearance Cl_H is the product of the blood flow and extraction ratio, it is also called the hepatic extraction flow, and its upper value is limited by the blood flow. Hence, units are volume·time^{-1} such as L·min^{-1}. The hepatic blood flow varies between 1–2 L·min^{-1} in humans and is on average 1.4 L·min^{-1}.

For illustration purposes, let us compare the extraction process of a compound across an organ with an assembly line (Figure 2.11). Assume that subject *A* in Figure 2.11 has the capacity of filling (*c.f.*, hepatic blood flow Q_H) 100 bottles per unit time and subject *B* of corking (*c.f.*, intrinsic clearance Cl_{int}) 1,000 bottles per unit time. The overall result will only be 100 filled and corked bottles per unit time because the rate-limiting step is the number of bottles that are actually filled per unit time. If subject *A* increases or decreases the production of filled bottles, the end-product will increase or decrease accordingly, since subject *B* has an over-capacity. This illustrates flow-dependent clearance.

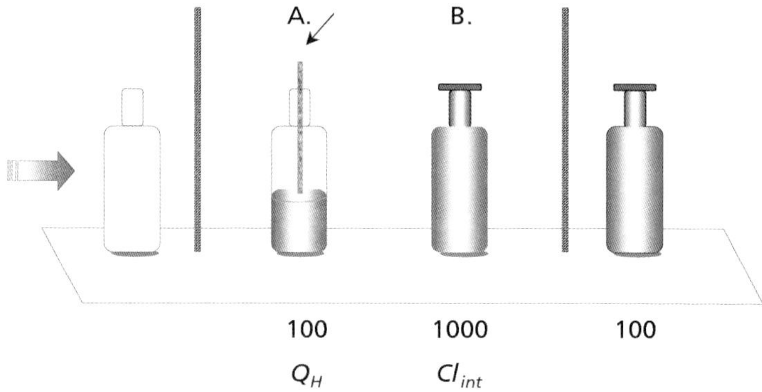

Figure 2.11
Flow dependent kinetics symbolized using of an assembly line as an analogy. Subject A is the rate limiting step, and will determine the number of end-products per unit time. If A increases or decreases filling capacity, the number of end-products will change accordingly.

Now assume that subject *A* in Figure 2.12 has the capacity of filling (*cf.* hepatic blood flow Q_H) 1,000 bottles per unit time and subject *B* of corking (*c.f.* intrinsic clearance Cl_{int}) only 100 bottles per unit time. The result will still only be 100 filled and corked bottles per unit time. If subject *A* increases or decreases his production of filled bottles, the end-product will not change, as long as subject *B* is the rate limiting step. However, if B increases its capacity of corking bottles, the end result of filled and corked bottles will also increase. This situation can be compared to flow independent clearance.

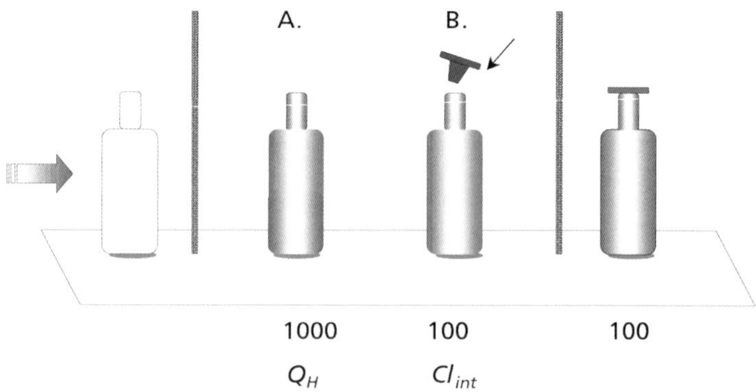

Figure 2.12
Kinetics dependent on the metabolic capacity symbolized using the assembly line analogy. Subject B is the rate limiting step, and will determine the number of end products per unit time. If B increases or decreases corking rate, the number of end-products will change accordingly.

Recall that clearance can either be viewed as a proportionality term between exposure (such as the plasma concentration-time curve *AUC*) and dose, or the rate of elimination and concentration. The area can either be presented as area from time zero to infinity, or as the area within a dosing interval τ from zero to τ. Clearance can also be assessed from a more physiological point of view as the product of the blood flow through the clearing organ and the extraction ratio. Figure 2.13 summarizes various methods to estimate and use clearance.

Clearance

Plasma $$Cl \cdot AUC^{24}_{0repeated} = Dose_{repeated}$$

Cl (vol/time)
Area (conc · time)
Dose (amount)

Plasma $$Cl = \frac{Dose_{repeated}}{AUC^{24}_{0repeated}} = \frac{Dose_{single}}{AUC^{\infty}_{0single}}$$

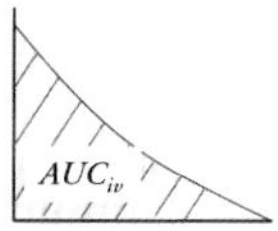

Physiological $$Cl = CL_H + CL_R = Q_H \cdot E_H + Q_R \cdot E_R$$

Cl (vol/time)
Q_H (vol/time)
E_H (-)

Figure 2.13
Relationship between clearance, area under the plasma concentration-time curve and dose within a dosing interval (24 h in this particular example) at steady-state (upper row). Relationship between clearance and the ratio of either dose at steady-state-to-area under the plasma concentration-time curve within a dosing interval, or the ratio of a single dose-to-area under the plasma concentration-time curve from time zero to infinity (middle row). The bottom row shows the additive effect of hepatic CL_H and renal CL_R clearance, which are also expressed as their perfusion rates and extraction ratios for the liver and kidney.

2.4.3 Elimination rate constant and half-life

Elimination rate constant K, also called the fractional rate constant, relates to the amount of drug in the body eliminated per unit time ($time^{-1}$). -*K* is the slope of the line resulting from a plot of ln(*C*) *versus* time (see derivation of $t_{1/2}$ in Figure 2.14). The biological half-life of the drug $t_{1/2}$ is the time required to reduce by half the amount of drug remaining in the body. *K* may also be directly estimated from the half-life $t_{1/2}$

$$K = \frac{ln(2)}{t_{1/2}} = \frac{CL}{V} \tag{2:9}$$

As evident from Equation 2:9, *K* is equal to clearance *Cl* divided by volume *V*. The *Cl*, *K*, and $t_{1/2}$ parameters relate to first-order processes. This means that they are dose- and concentration-independent. If the dose doubles there is a similar doubling of the plasma concentration. The slope –*K*, clearance (*Dose*/*AUC*) or $t_{1/2}$ will not change. Note that *Cl* and *V* are the independent variables and *K* and $t_{1/2}$ are functions of *Cl* and *V*. Half-lives are dependent on both elimination *Cl* and distribution *V*. Since *Cl* and *V* are the primary (and independent of each other) pharmacokinetic parameters of interest, they are often used instead of *K*. Clearance *Cl* and volume *V* are often referred to as primary or estimated parameters, whereas half-life $t_{1/2}$, elimination rate constant *K* or area under the plasma concentration-time curve *AUC* ($mass \cdot vol^{-1} \cdot time$, *e.g.*, $\mu g \cdot L^{-1} \cdot h$) are called secondary or derived parameters. The relationship between clearance, volume and half-life is shown in Figure 2.14.

Figure 2.14
Derivation of the relationship between the two primary pharmacokinetic parameters clearance *Cl* and volume of distribution *V*, in relation to the secondary pharmacokinetic parameter biological half-life $t_{1/2}$. Ln(2) signifies the natural logarithm of 2 (which is approximately 0.693).

Half-life is determined by *Cl* and *V*

$$C = C^0 \cdot e^{-\frac{Cl}{V}\cdot t} \longrightarrow C_{50} = C_{100} \cdot e^{-\frac{Cl}{V}\cdot t_{1/2}} \longrightarrow \frac{C_{50}}{C_{100}} = e^{-\frac{Cl}{V}\cdot t_{1/2}}$$

$$\longrightarrow ln\left(\frac{C_{50}}{C_{100}}\right) = ln\left(e^{-\frac{Cl}{V}\cdot t_{1/2}}\right) \longrightarrow ln\left(\frac{C_{50}}{C_{100}}\right) = ln\left(\frac{1}{2}\right) = -\frac{Cl}{V}\cdot t_{1/2}$$

$$\longrightarrow ln(2) = \frac{Cl}{V}\cdot t_{1/2} \longrightarrow t_{1/2} = ln(2)\cdot\frac{V}{Cl}$$

Most computer programs will present data in log-linear plots as shown in Figure 2.9. To estimate *K*, one would perform a regression analysis on the ln(*C*) versus time; the resultant slope is $-K$ and intercept is ln(C^0), which are the model parameters in this case. Initial estimates of *K* and $t_{1/2}$ can be derived from the slope. *V* is estimated from the intravenous dose D_{iv} and the intercept C^0. Using the data in Figure 2.9 as an example

$$slope = \frac{y_2 - y_1}{x_2 - x_1} = -K = \frac{ln(1000) - ln(40)}{0 - 150} = -0.022\ min^{-1} \quad (2:10)$$

The half-life becomes

$$t_{1/2} = \frac{ln(2)\cdot V}{Cl} = \frac{ln(2)}{K} = \frac{0.693}{0.022} \approx 32\ min \quad (2:11)$$

The intercept of the concentration axis is 1000 µg·L^{-1} after an intravenous dose of 10000 µg. According to Table 2.1, the volume of distribution *V* can be calculated as

$$V = \frac{Dose}{C^0} = \frac{10000}{1000} = 10\ \text{L} \quad (2:12)$$

The total area under the curve, $AUC_{0-\infty}$ is either obtained by direct integration of Equation 2:4 from zero to infinity

$$AUC_0^\infty = \int_0^\infty C^0 e^{-K\cdot t} dt = C^0\left[\frac{e^{-K\cdot\infty}}{-K} - \frac{e^{-K\cdot 0}}{-K}\right] =$$
$$= C^0\left[\frac{0}{-K} - \frac{1}{-K}\right] = \frac{C^0}{K} = \frac{1000}{0.022} \approx 47{,}000\ \frac{\mu g}{L}\cdot min \quad (2:13)$$

Clearance is estimated from dose D_{iv} divided by the total area under the curve $AUC_{0\text{-}\infty}$

$$Cl = \frac{D_{iv}}{AUC_0^{\infty}} = \frac{D_{iv}}{\left[\frac{C^0}{K}\right]} = 0.22\,L \cdot min^{-1} \qquad (2:14)$$

Three other example drugs and their pharmacokinetic properties after a single rapid intravenous injection, are shown in Figure 2.15. They have similar half-lives although their concentration-time courses differ substantially after the same dose. Warfarin has a low volume of distribution giving it the highest initial plasma concentration. Of the three, warfarin also has the lowest clearance, resulting in the largest area under the plasma concentration-time curve *AUC* (*Dose*/*Cl* = *AUC*). Nortriptyline has a large volume of distribution giving a low initial concentration, and a high plasma clearance resulting in a low *AUC*. It is therefore apparent that the half-life will not indicate the extent of the maximum plasma concentration such as C^0, or the area under the plasma concentration-time curve. A half-life only tells how quickly the plasma concentration declines.

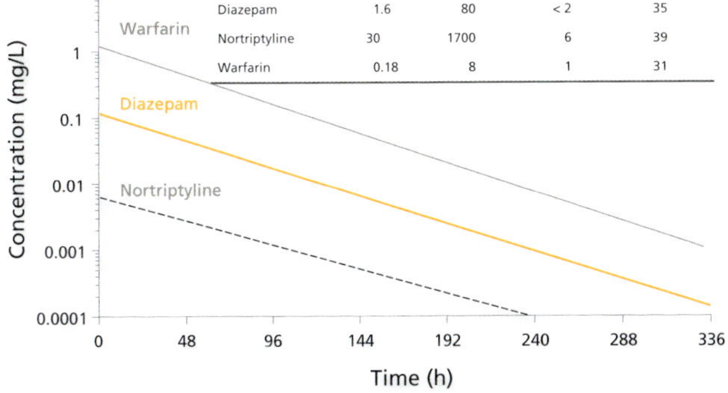

Compound	Clearance ($L \cdot h^{-1}$)	Volume (L)	Free fraction (%)	Half-life (h)
Diazepam	1.6	80	< 2	35
Nortriptyline	30	1700	6	39
Warfarin	0.18	8	1	31

Figure 2.15
Semi-logarithmic plot of warfarin, diazepam and nortriptyline plasma concentration-time data following an intravenous bolus dose equivalent to 10 mg. The data show mono-exponential decline curves described by Equation 2:4. Note the similar half-lives (30–40 min) in spite of a 200-fold difference in the clearances and volumes for the three compounds.

The half-life is therefore of limited value also from a pharmacodynamic point of view, unless put into an exposure (C_{max}, C_{min}, EC_{50}) context. Clearance, on the other hand, not only determines *AUC* and the steady-state concentration, but also the dose needed to establish a particular target effect when combined with effective plasma concentrations. How clearance determines the steady-state plasma concentration will be discussed in the next section. Remember that clearance governs *AUC* and average concentration, and therefore influences the pharmacological response.

Bear in mind also that a rapid bolus injection may not be the most appropriate way of designing a pharmacology experiment because the response changes (increases or decreases) as a function of the falling plasma concentration. Ideally, the increase and decrease in response should be a function of both increasing and decreasing plasma concentrations. A rapid drug input rate with high C_{max}, such as after an intravenous bolus, may also increase the risk of adverse reactions. In Section 2.5 we show how pharmacological experiments can be optimized by means of constant-rate infusions.

2.5 Constant rate infusion

So far only the disposition of a drug after intravenous bolus dosing has been characterized. Under certain circumstances it may be more feasible to give a constant intravenous infusion for studying the pharmacological effect or the disposition of a drug. The onset, intensity and duration of response are more easily observed and handled when a drug is given as an intravenous infusion, *i.e.,* under controlled conditions. Measuring drug response during and after a constant-rate infusion allows for monitoring of pharmacodynamic alterations both as plasma concentration rises and falls.

If a drug is infused at a constant rate over a period of time, the plasma concentration will rise as shown in Figure 2.16. Fifty percent of steady state is reached after one half-life, 75 % after 2 half-lives, and 87.5 % after 3 half-lives. From a practical point of view, therefore 90 % of steady state is considered to have been attained within 3–4 half-lives.

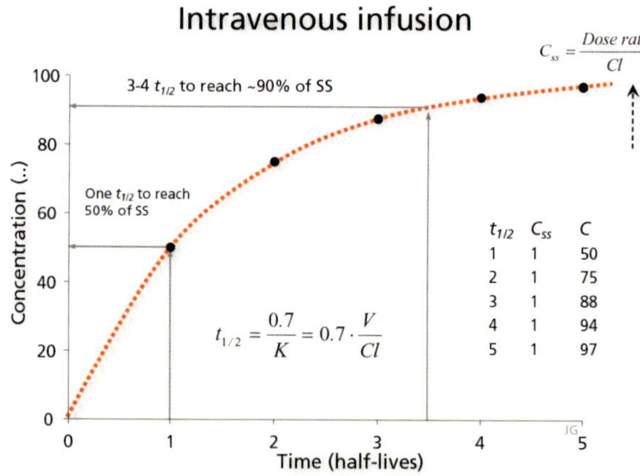

Figure 2.16
The plasma concentration-time course following a constant rate infusion. Note that the concentration is plotted against half-lives of the drug. Infusion during one half-life establishes 50 % of steady state (SS). Three to four half-lives are needed to reach about 90 % of the steady state concentration.

The relationship between the concentration C and time during an infusion like in Fig. 2.16 may be expressed mathematically for a one-compartment model by means of Equation 2:12 during the infusion.

$$C = \frac{R_{in}}{Cl} \cdot \left[1 - e^{-\frac{Cl}{V} \cdot t} \right] \qquad (2:15)$$

where R_{in} denotes the rate of infusion. The steady state concentration C_{ss} can be calculated as R_{in} divided by Cl, hence

$$C_{ss} = \frac{R_{in}}{Cl} \qquad (2:16)$$

Although the time to steady-state is governed by volume and clearance via the half-life, clearance is the sole parameter that determines the steady-state level and maintainance dose required to keep the target plasma concentration at steady state. A typical experimental set-up of a preclinical cardiovascular study is shown in Figure 2.17. Note that both plasma concentration and response rises and falls during and after cessation of infusion. The concentration-response plot (Figure 2.17, lower plot) therefore displays a counter clockwise loop. This will be further discussed in Chapter 6, addressing temporal differences between plasma concentration and response. Notice that data from an acute experiment with observed temporal differences, are used to predict the concentration-response conditions at equilibrium. That is the purpose of modelling.

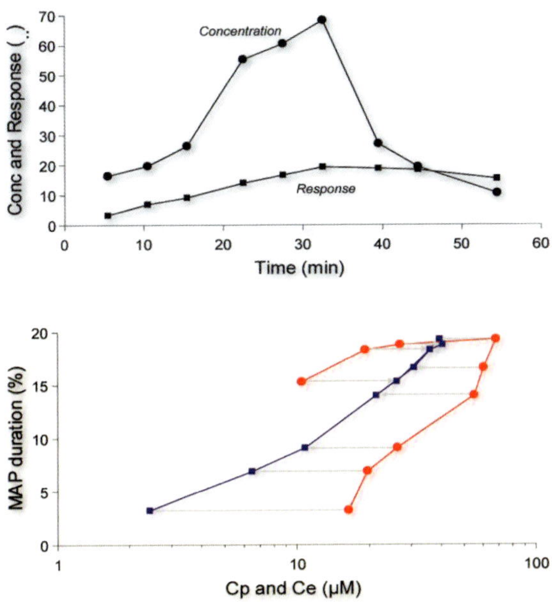

Figure 2.17
Upper: Concentration- and response-time data of compound X after two consecutive intravenous infusions followed by washout to a dog.
Lower: Concentration-response data from the same experiment plotted in time order (red data) showing the disequilibrium between plasma exposure and response. The blue curve represents the equilibrium concentration-response relationship obtained from modeling response-time data.

2.6 Integration of clearance and volume

So far, two ways to study the disposition of a drug have been examined, namely bolus and constant rate intravenous infusion. It is now timely to systematically integrate knowledge about *Cl* and *V* and how they affect the concentration time course after an intravenous bolus dose or constant rate intravenous infusion. The impact of a change in clearance and volume on the plasma concentration time course after an intravenous bolus dose is illustrated in Figure 2.18.

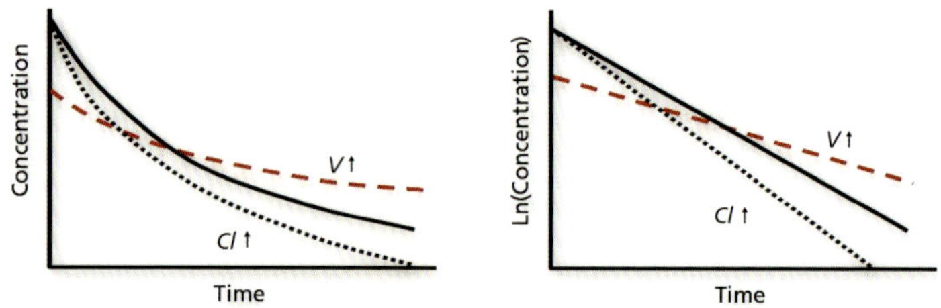

Figure 2.18
Linear (left) and semi-logarithmic (right) plot of the concentration-time curve of a one-compartment drug after an acute intravenous bolus dose. The standard situation is shown by the solid black line, an increase in clearance by the dotted black line and an increase in volume of distribution by the dashed red line.

There may also be safety reasons why a rapid bolus injection or oral administration is not the optimal mode of administration. If during an intravenous infusion adverse reactions are observed, the infusion can be stopped immediately and the plasma concentration allowed to fall, thereby hopefully also resolving any untoward effect. A controlled build-up of exposure via constant intravenous infusion regimens in preclinical pharmacological experimentation is thus also preferable from an ethical point-of-view, as it allows for simultaneous extraction of valuable PK and PD information without compromising animal welfare and, in turn, quality of data. Clearly, interrupting exposure build-up after bolus injections or oral administration modes is not feasible.

The impact of either a change in the infusion rate or a change in clearance on the time course of a drug in plasma is shown in Figure 2.19. An increase or decrease in the infusion rate affects the steady state concentration but not the time to steady state, provided the kinetics is linear. However, an increase or decrease in plasma clearance not only changes the steady state level of drug but also the time to steady state. An increase in clearance via enzyme induction decreases the time to steady state since the half-life is decreased. A decrease in clearance has the opposite effect in that it prolongs the half-life and hence time to steady state will increase. Changes in clearance without corresponding adjustments in dosing rate/interval may therefore have significant impact also on pharmacodynamic outcome of drug treatment. For example, an increased clearance of an antibiotic medication may result in loss of antibacterial effect, potentially paving the ground for increased bacterial resistance development and sustained infection. Conversely, a decreased clearance may result in too high drug exposure, with enhanced risk for undesired adverse or toxic reactions, such as for instance in the case of lithium treatment of bipolar disorder.

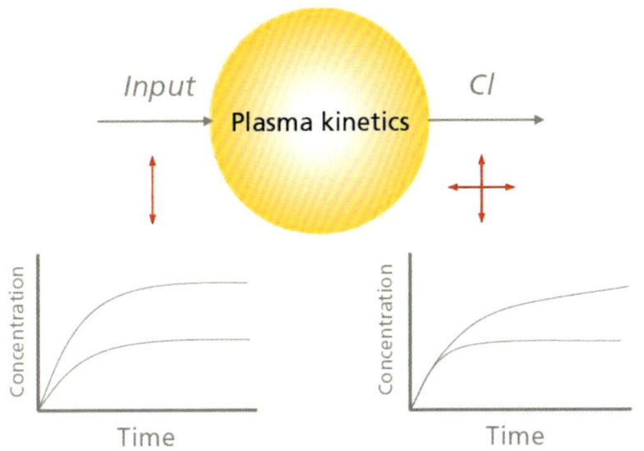

Figure 2.19
Schematic illustration of a change in the infusion rate which has no effect on the time to steady state, but does change the steady state level (left hand panel). A change in clearance affects not only the steady state level but also the time to steady state (right hand panel). An increase in clearance decreases time to steady state whereas a decrease in clearance increases time to steady state.

In the next example we combine pharmacokinetic information obtained after intravenous bolus administration with the expected outcome when drug is given as a constant intravenous infusion regimen. The diagram in Figure 2.20 shows how equal volumes and different clearances (A, B), equal clearances and different volumes (C, D), and different clearances and different volumes (E, F) affect the time course of drug in plasma after either intravenous bolus dosing (equal doses, upper row of graphs) or constant rate intravenous infusion (equal dosing rates, lower row of graphs). Note that the ratio of volume-to-clearance is a constant in the last example (E, F). Consequently, E and F have equal half-lives and a parallel fall in the plasma concentrations following a bolus dose. The bolus doses and intravenous infusion rates are the same for all cases.

The upper row of concentration-time graphs shows how clearance *Cl* determines the area under the curve *AUC*, and how volume *V* determines the peak concentration C^0. The larger the clearance the lower the *AUC*, while the larger the volume, the lower the maximum plasma concentration C^0, and *vice versa*. Since A and B have similar volumes, their maximum plasma concentrations C^0 will be the same. However, clearance is larger for B than for A. Hence, AUC_A is greater than AUC_B.

Examples C and D have similar clearance and therefore their *AUC* values will be the same. However, *V* of D is less than *V* of C and therefore C^0 of D will be higher. There is less space in D for drug to distribute into, so the peak concentration will be higher. In the last example, E and F differ in both clearance and volume, but their volume-to-clearance *V/Cl* ratios are the same, which is also observed from their identical half-lives. Since clearance for F is greater than for E, AUC_F will be smaller than AUC_E. Also, since E has the smallest volume, the maximum plasma concentration will be higher.

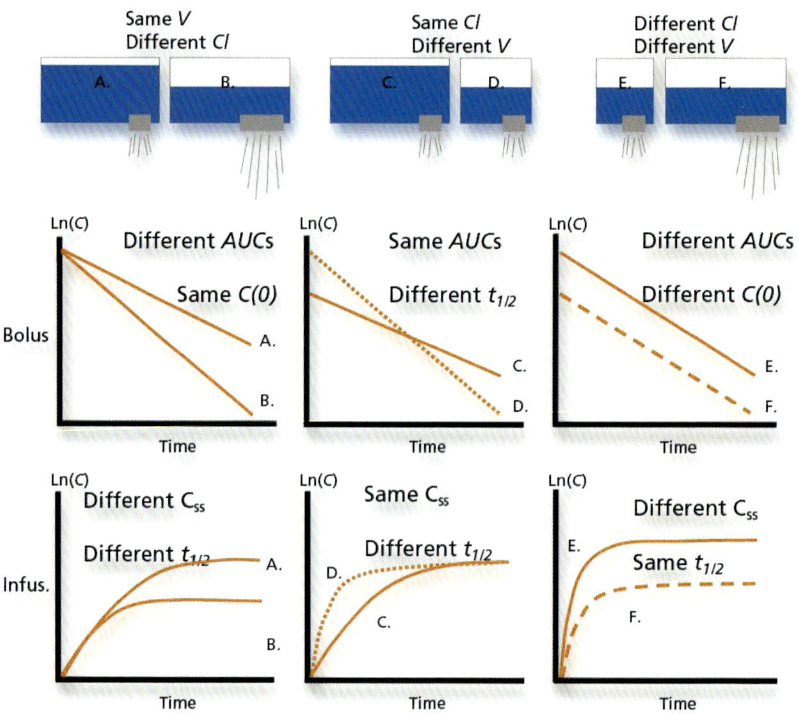

$$t_{1/2} = \frac{0.7}{K} = 0.7 \cdot \frac{V}{Cl}$$

$$C_{ss} = \frac{Dose\ rate}{Cl}$$

Figure 2.20
This figure illustrates the kinetics of six different systems. It compares how equal volumes but different clearances (left column), equal clearances and different volumes (middle column), and different clearances and different volumes but equal half-lives (right column), affect the time course of drug in plasma after bolus dosing and constant rate intravenous infusion.

The lower row of plots in Figure 2.20 displays the concentration time courses after a constant rate infusion. It is assumed that the dosing rate is the same. Note that only clearance affects the steady state concentration, but both clearance and volume determine the time to steady state via the half-life. Steady state is reached after 3 to 4 half-lives. Case A has the lower clearance and will therefore establish a higher steady state concentration C_{ss}. The time to steady state is longer for A since the half-life is greater than for B. C and D have the same clearance and will also have the same C_{ss}. Since $t_{1/2}$ is greater for C than D, the time to steady state will be longer. The lower clearance for E causes the higher C_{ss}. The time to steady state will be the same for both E and F since their half-lives are the same.

When a drug is given repeatedly as intravenous bolus doses one gets a concentration-time course as shown in Figure 2.21. Steady-state is reached when two consecutive concentration-time courses are the same. Maximum plasma concentrations C_{max} and trough plasma concentration C_{min} and Area Under the plasma concentration-time Curve $AUC_{0\text{-}tau}$ will be the same at each dosing interval at steady-state. This figure also illustrates the fact that the total area under the drug plasma concentration-time course from zero to infinity equals the area under the plasma concentration-time curve within a dosing interval at steady-state. The expression 'integral' is sometimes used as a substitute for the area under the plasma concentration-time curve. The area has a unit concentration·time (μg·L^{-1}·h or nmol·L^{-1}·h).

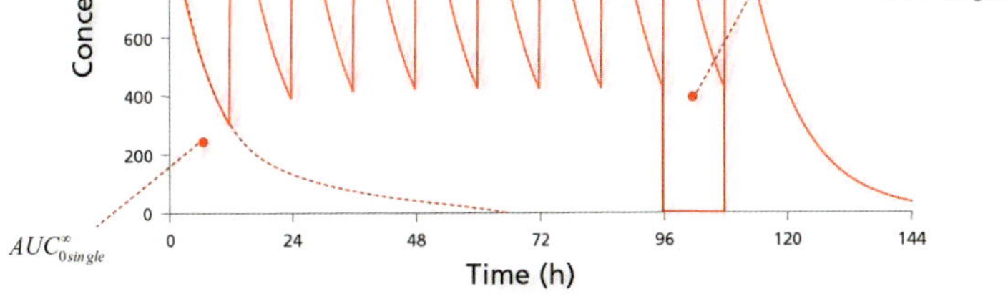

Figure 2.21
Concentration-time profile of repeated bolus injections with a dosing interval of 12 h. The area under the the plasma concentration-time curve from zero to infinity after a single dose $AUC_{0\text{-}inf(single)}$ equals the area within a dosing interval 0–12 h at steady-state. Steady-state implies that the total amount eliminated within a dosing interval equals the dose. The areas within two consecutive dosing intervals are the same at steady-state.

2.7 Extravascular administration

All routes of drug administration other than directly into the blood circulation (intravenous and intraarterial) are referred to as extravascular administration. Oral, subcutaneous (under the skin), dermal (onto the skin) and pulmonary (inhaled via the lungs) dosing are examples of the extravascular route. They are used both clinically and in animal experimental models. The administration of a drug by any route other than systemic (intravenous) will introduce an absorption step. The change in body drug content or plasma drug concentration will now be more complex, since the rate of change of drug will be a function of both the absorption rate and the elimination rate. When a drug has reached the blood plasma pool it is further transported to various tissues and organ systems. The drug eventually reaches the target site. Most target sites are located on a cell membrane, inside the cell or its nucleus or are circulating targets in the blood. Figure 2.22 illustrates additional distributional processes prior to target binding or before target activation is being established. The drug may be actively transported into or out of the cell, passively diffuse or act upon transmembrane (TM) receptors. In the latter case the drug molecule never needs to pass into the cell to elicit its pharmacological action. *Prior* to arrival at its actual therapeutic cellular target or target tissue, a drug may also face other absorption and distribution hurdles. Examples of such include, *e.g.*, gastrointestinal, blood-brain and placental barriers, the permeabilities through which depend on drug physicochemical *vs.* membraneous properties and/or access to transporters. From a common PK perspective, cellular target and barrier processes described are however treated in a similar manner.

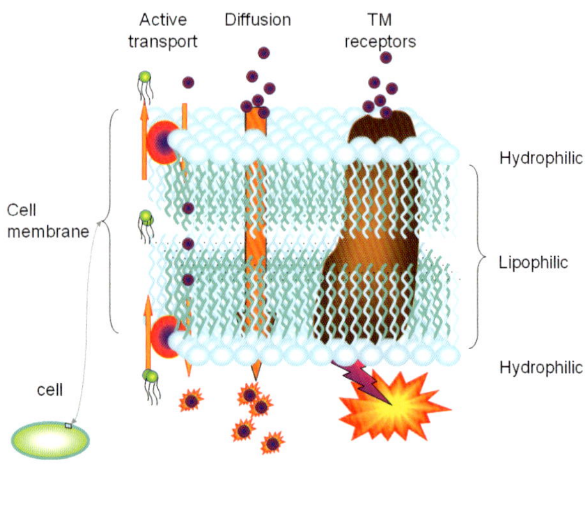

Figure 2.22
Schematic illustration of a cell membrane. Access of a molecule into the intracellular space can occur via passive diffusion (middle arrow), transporters (left hand side transport) or via binding to transmembrane receptors. In the first two cases the drug molecule has to be physically present intracellularly, whereas when the drug molecule binds to to the transcellular membrane protein its pharmacological action is initiated from outside the cell and further transduced by the intracellular portion of the transmembrane (TM) receptor protein.

A typical plasma concentration-time plot after extravascular administration of an instant release IR dosage form or a solution is shown in Figure 2.23. After administration of an oral dose at time zero the plasma concentration increases because rate of absorption (input of drug into the system or plasma) is greater than the rate of elimination (loss via excretion of drug into urine and/or metabolism). At the peak concentration rate of absorption equals the rate of elimination, and then the latter dominates post-peak.

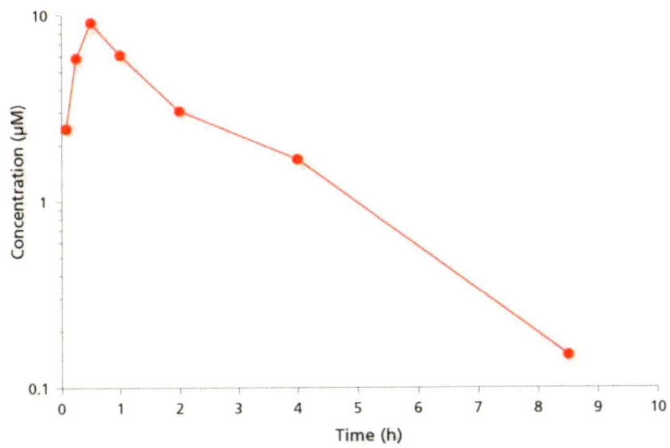

Figure 2.23
Semi-logarithmic plot of plasma concentration-time data following an extra-vascular dose equivalent to 10 µmol·kg^{-1} of a standard instant release dosage form.

A controlled release dosage form means a matrix of drug substance combined with pharmacologically inert material that integrates or dissolves more slowly than a conventional tablet. The dosage form controls the release of drug substance. The rate and extent of drug substance release from the dosage form can then control the *onset*, *intensity* and *duration* of the pharmacological effect. Data are obtained after administration of a controlled-release dosage form (Figure 2.24). Note how late (at about 6 to 8 h) the peak concentration occurs in this example. The terminal half-life of approximately 8 h is due to absorption rate-limited elimination, which will result in a small fluctuation between C_{max} and C_{trough}. The lag-time corrected concentration-time course (Equation 2:19) is represented by the red line.

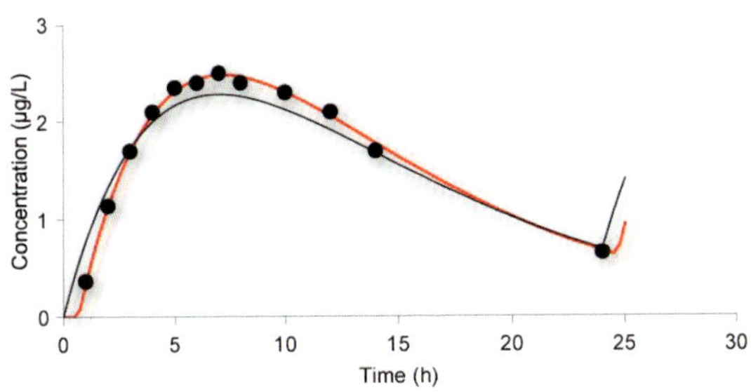

Figure 2.24
Semi-logarithmic plot of plasma concentration-time data following an extra-vascular dose equivalent to 100 µg of a controlled release CR dosage form. The data show mono-exponential decline. The red line includes a lag-time. Black line has no lag-time.

The function for an orally administered drug that obeys one-compartment kinetics will take the following form

$$C = \frac{K_a \cdot FD_{po}}{V(K_a - K)}\left[e^{-K\cdot t} - e^{-K_a\cdot t}\right] \tag{2:17}$$

where K_a denotes the apparent first-order absorption rate constant and F denotes the bioavailability or extent of absorption. The bioavailability is the fraction of the dose that reaches the systemic circulation intact. The bioavailability is set to 100 % after intravenous dosing, whereas F is less than or equal to 100 % after extravascular dosing. The behaviour of Equation 2:17 is shown by the grey curve in Figure 2.25. This schematic picture delineates the contributing parts to a concentration-time profile obtained after extravascular dosing.

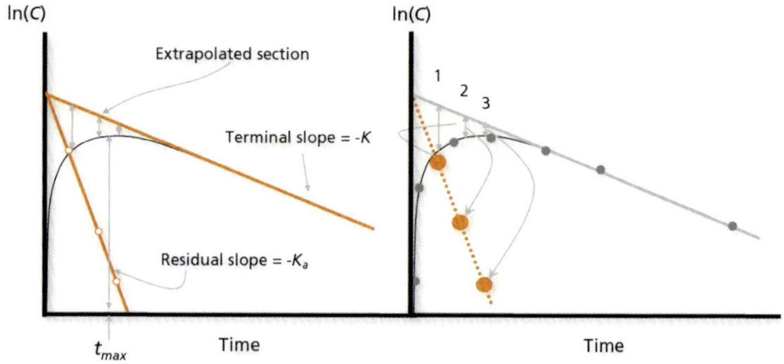

Figure 2.25
Semi-logarithmic plot of concentration-time data of compound X after oral administration of 20 mg. Note the interception of the back-extrapolated terminal slope and the residual (red) slope at time zero. The numbers 1, 2, and 3, depict arbitrarily chosen intervals used to extract the differences between extrapolated and observed curve, and using these data to construct the residuals plot (*cf.* curved arrows in righ-hand graph).

The slope of the linear portion of the terminal phase of the curve when plotted on a semi-ln (ln denotes the natural logarithm) scale is equal to $-K$, assuming that the absorption is much more rapid than elimination ($K_a >> K$). The resulting dotted time course in Figure 2.25 (right graph) is called the residual. This difference $C_{residual} = C_{extrapolated} - C_{observed}$ is then plotted in the same figure and a straight line is drawn through the data. The slope of the linear residual portion of the initial phase of the curve is equal to $-K_a$. The rate constants K_a and K may be determined by the curve stripping method, also called the method of residuals (see also Figures 2.25 and 2.26). The area under the plasma concentration-time curve can be obtained by integrating Equation 2:17 to yield

$$AUC_0^{\infty} = \frac{FD_{po}}{VK} = \frac{FD_{po}}{Cl} \tag{2:18}$$

The time course of the concentration in plasma of certain compounds suggests a time delay between oral administration and the apparent onset of absorption (Figure 2.26). The dissolution of the drug particles and the absorption of dissolved compound into the systemic circulation takes a certain amount of time until the concentration of the drug is measurable in plasma (Equation 2:19).

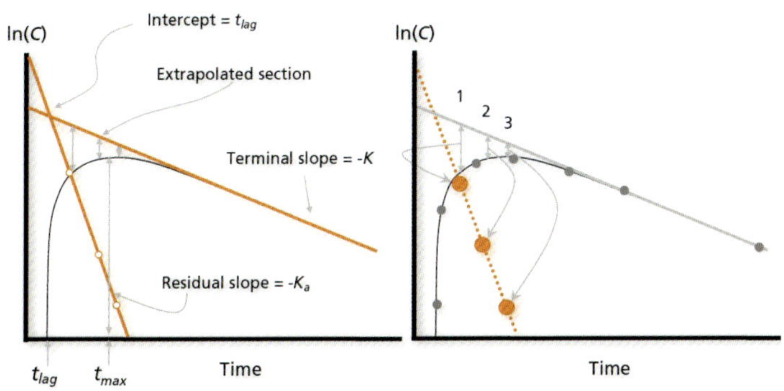

Figure 2.26
Semi-logarithmic plot of concentration-time data of compound X after oral administration of 20 mg. Note the interception of the back-extrapolated terminal slope and the residual (red) slope at about 15 min. The experimental data are shifted to the right in time in comparison to Figure 2.25 due to the lag time.

The analysis is started by estimating the rate constants of the absorption and elimination phase. The rate constants are obtained from the slopes

$$C = \frac{K_a \cdot FD_{po}}{V \cdot (K_a - K)} \cdot \left[e^{-K\cdot(t-t_{lag})} - e^{-K_a\cdot(t-t_{lag})}\right] \tag{2:19}$$

It can be seen that the back-extrapolated terminal slope and the absorption slope intersect at t_{lag} (Figure 2.26). This suggests that there is a lag-time involved before drug appears at the sampling site. However, note that t_{lag} is usually directly obtained from a plot of experimental data, including the back-extrapolated and residual slopes.

The consequences of ignoring a lag-time after oral dosing are shown in Figure 2.27. Both the maximum concentration C_{max} and the the corresponding time t_{max} are biased when the lag-time t_{lag} is ignored. In this particular case the total area under the curve is not dramatically changed, but from a biopharmaceutical characterization point of view, the rate of absorption will give erroneous results. Again, using the wrong pharmacokinetic model to 'drive' the pharmacodynamics will then give an erroneous *onset*, *intensity* and *duration* of pharmacological action.

Figure 2.27
Concentration-time data of test compound X. The solid line is the lag-time model (inset, Equation 2:19) and the dotted curve is lacking a lag-time parameter (Equation 2:17). The vertical dashed red line shows the predicted time of the peak concentration.

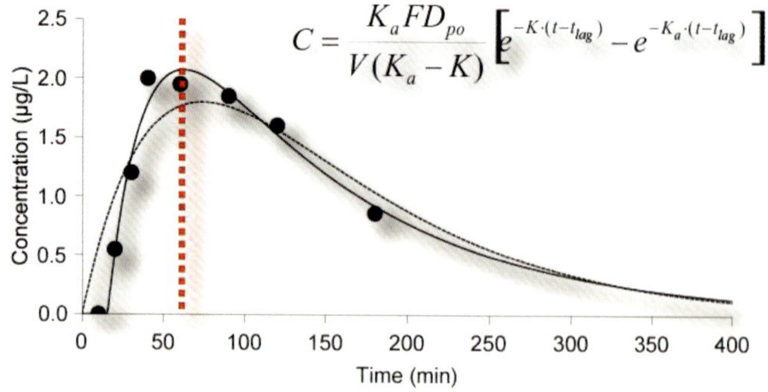

An example of zero order drug absorption is shown in Figure 2.28. This means that the dose is absorbed at a constant rate up to about 4–5 h (horizontal orange bar), after which no more drug is left in the gut to be absorbed. A first-order absorption model with lag-time was also fitted to the data, but as can be seen, there is a systematic deviation between experimental and model predicted data. The first-order absorption model both misses the maximum concentration and is inconsistent with terminal plasma concentrations.

Figure 2.28
Concentration-time data of drug Y. The solid discontinuous line abruptly shifting direction at 4.5 h is the zero-order absorption model and the smooth curve is the first-order absorption model with lag-time. The horizontal orange bar shows the predicted zero-order absorption time of the former model.

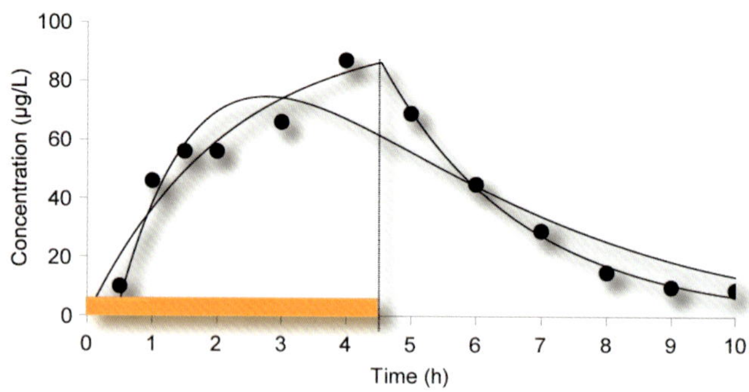

An example of controlled drug delivery is the zero order absorption of nicotine across the skin from a nicotine patch (Figure 2.29). Nicotine is initially (for the first 6 h) rapidly released from the patch. The rate limited release membrane system then governs delivery of nicotine to the body between 6–16 h. The patch is removed at 16 h but the plasma nicotine concentration is still maintained for a further hour before starting to decline. This is probably because a depot of nicotine has been built up in the skin.

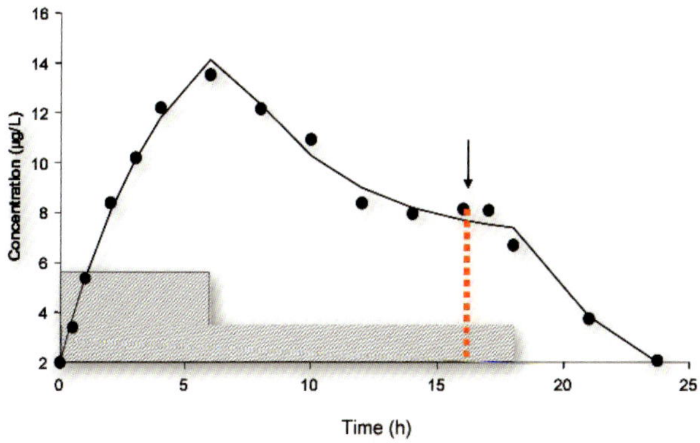

Figure 2.29
Concentration-time data of nicotine after application of a transdermal therapeutic delivery system – transdermal patch. The solid discontinuous line is the model predicted concentration-time course and includes an initial burst of nicotine release during 6 hours followed by a maintenance nicotine release up to 17 hours. The vertical dashed red line shows the time of removal of the nicotine patch. Note that there is a 1–2 hour delay until the terminal decline in plasma nicotine concentration. The delay time may be due to the remaining dermal depot of nicotine.

Typically, the elimination rate constant would be estimated from the terminal slope of the extravascular concentration-time profile, for example after ingestion of an instant release tablet. In some cases, however, the absorption process will be the rate limiting step, as with a controlled release tablet. This is known as the flip-flop situation, where the ascending limb of the concentration-time curve will contain information about the elimination rate constant and the descending or terminal portion will contain information about the absorption rate constant. In order to discriminate between the two phases, the true elimination phase of the drug after an intravenous dose needs to be calculated. That is shown for the flip-flop situation in Figure 2.30. Note that in the right hand figure the slope of the terminal portion of the extravascular curve is not parallel to the slope of the intravenous curve. The latter slope gives the true elimination rate constant and the slope of the former gives the absorption rate constant.

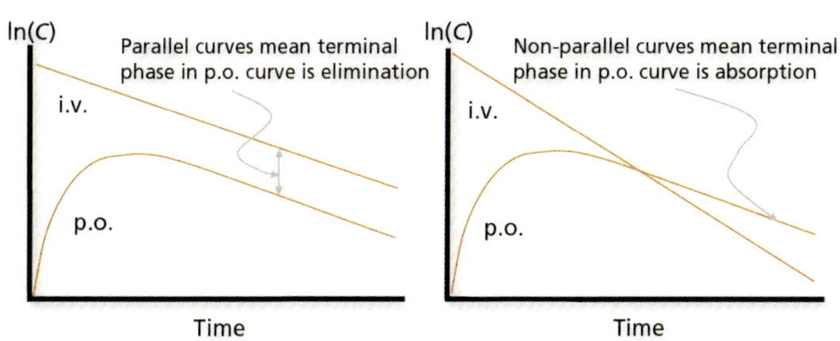

Figure 2.30
Plasma concentration-time data following an extravascular dose. The left hand figure demonstrates a system where the elimination rate constant is obtained from the terminal portion of the oral curve since it declines in parallel with the intravenous curve. The intravenous curve represents only elimination. The right hand figure demonstrates the flip-flop situation for a system that displays absorption rate limited elimination. Therefore, the terminal portion of the oral curve will not run in parallel with the intravenous curve. The latter reflects the true elimination of drug, whereas the elimination in the terminal phase of the oral curve is determined by absorption.

Figure 2.31 compares three different scenarios of the concentration-time course of an oral dose and a matched constant intravenous infusion followed by washout. The peak concentration after the oral dose and 90 % of steady-state following intravenous dosing are indicated by a red bar. An increase in hepatic intrinsic clearance from scenario 1 to 2 lowers the area under the two concentration-time courses since clearance increases. Half-life is decreased and the peak time and time at 90 % of steady-state are also shortened as indicated by the leftward shift of the red bars. Note that the half-life after oral dosing is slightly longer than indicated by the intravenous infusion washout curve. The reason for this is that absorption-rate limited elimination occurs in the oral curve. When intrinsic clearance is increased further from scenario 2 to 3, the area under the oral plasma concentration-time curve decreases, peak concentration is slightly shifted to the left, but the terminal phase is still absorption rate limited. The area under the curve and steady-state concentration are inversely proportional to intrinsic clearance after oral dosing. If intrinsic clearance is increased, the area will decrease (scenario 1–3a). However, in the case of the intravenous infusion, an increase in intrinsic clearance decreases the plasma concentration until clearance becomes perfusion limited. This means when the value of clearance approaches the blood flow through the eliminating organ. There is no further decrease in the plasma concentration when clearance have become perfusion limited, which can be seen in scenario 2b–3b. The plasma concentration-time course after the intravenous infusion regimen is basically unaffected. The reason for this is that clearance of drug has become perfusion rate limited. Time to steady-state and the actual steady-state plasma concentration do not change (decrease) further. The reason for the differences between the oral and intravenous data is that the decrease in the oral *AUC* is due to the first-pass effect that is directly proportional to hepatic capacity, namely intrinsic clearance.

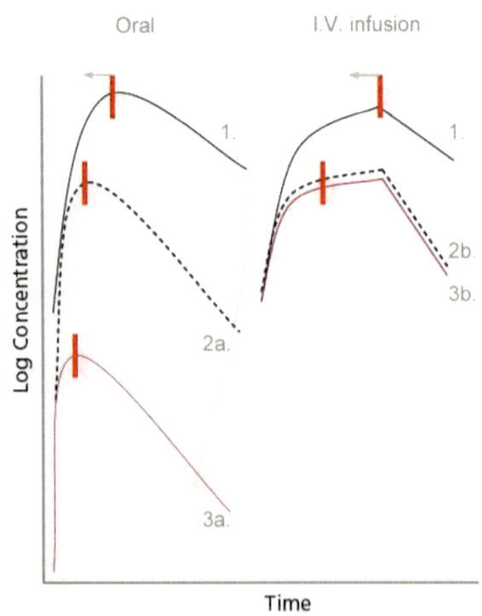

Figure 2.31
Simulated concentration-time course of oral and constant rate intravenous dosing plotted on a semi-logarithmic scale. Scenarios 1 to 3 are obtained by increasing the hepatic intrinsic clearance.

2.8 Extent of absorption - bioavailability

When a drug is given via the oral, subcutaneous or transdermal routes, only a fraction of the dose reaches the blood circulation. This fraction is a measure of the *extent* of absorption. The conventional way of estimating the *extent* of absorption, also called the bioavailability F, is by sequential dosing of the intravenous (systemic) and extravascular (oral) doses generally with an interval of either one day (in small animals) or one week (humans) between administrations. One reason for sequential dosing in the same subjects, rather than parallel dosing in different subjects, is a desire to minimize the influence of inter-individual variability in pharmacokinetics. Often, the intravenous dose is given first and plasma samples are collected. Subsequently the oral dose is given and plasma samples are again collected. The order of the intravenous and oral doses may be randomized. For estimation of bioavailability, the underlying assumption is that clearance is constant between the two occasions (remember though that some compounds induce or inhibit liver enzymes, which indirectly affects the concentration-time course of a second dose by means of a change in clearance). If however clearance remains constant, the dose can be used, divided by the area under the curve obtained from oral dosing ($Dose_{po}/AUC_{po}$) and intravenous dosing ($Dose_{iv}/AUC_{iv}$). The ratio of Cl_{iv}-to-Cl_{po} gives the bioavailability F (Figure 2.32).

$$Cl = \frac{Dose_{iv}}{AUC_{0iv}^{\infty}} = \frac{F \cdot Dose_{po}}{AUC_{0po}^{\infty}}$$

$$F = \frac{AUC_{0po}^{\infty}}{AUC_{0iv}^{\infty}} \cdot \frac{Dose_{iv}}{Dose_{po}}$$

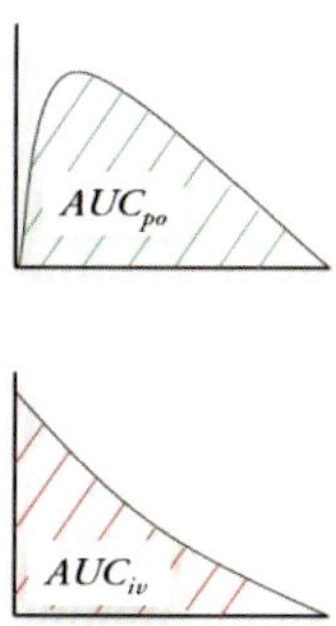

Figure 2.32
Summary of the principles used for estimation of plasma clearance and oral bioavailability. AUC_{iv} and AUC_{po} denote the area under the plasma concentration-time curves from zero to infinity after intravenous and oral administration of drug.

The magnitude of variability in the *rate* and *extent* of absorption between different compounds, routes of administration, individuals, species and strains, needs to be built into the assessment of a pharmacological effect. Since the same amount of compound given via different routes of administration will result in totally different exposure profiles, it is always necessary to relate effect measurements to plasma exposure. Figure 2.33 shows a semilogarithmic plot of one and the same compound, given at the same dose, but via different routes of administration. Intravenous and subcutaneous dosing demonstrate very little between-animal variability for this compound, whereas the intraperitoneal route has substantial between-animal variability in the plasma exposure to the compound. The latter may be explained by the fact that in three of the animals, the dose was inadvertently placed into the intestinal tract, resulting in a high and variable first pass loss during the absorption.

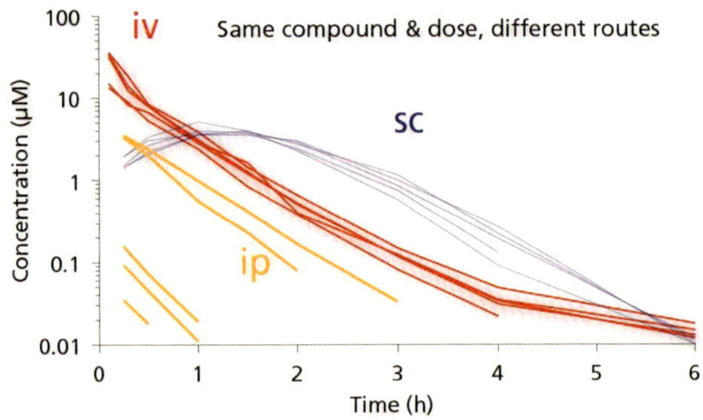

Figure 2.33
Semi-logarithmic concentration-time plot of compound X. The same compound and dose was administered via intravenous (iv), subcutaneous (sc) and intraperitoneal (ip) routes.

Another typical pharmacological experiment with test compounds A and B is shown in Figure 2.34. It shows a dose-response graph (left) and concentration-response graph (right) of compounds A and B after oral administration at different doses. The oral bioavailability of compound A is more than 95 %, whereas for compound B less than 5 %. The left hand graph gives the impression that compound A is more potent than compound B, a difference that can simply be explained by different bioavailabilities of A and B. When a concentration scale is used it can be seen that both compounds exhibit similar potencies of about 20 nM.

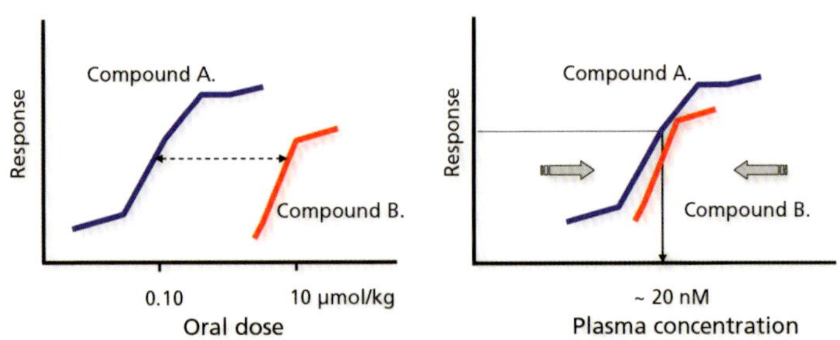

Figure 2.34
Semi-logarithmic dose-response (left) and concentration-response (right) plots of compounds A and B in an *in vivo* rabbit model. The bioavailability of compounds A and B are >95 % and <5 %, respectively.

This Section has covered reasons for why simultaneous measuring of exposure and pharmacological response is necessary when different routes, rates or compounds are considered. The *rate* and *extent* of absorption may vary extensively. The next Section covers how drug exposure in plasma varies after repeated or chronic administration and how drug accumulation can be assessed and what implications that may have on the pharmacological response.

2.9 Accumulation and time to steady-state

Chronic indications require chronic dosing. Many pharmacological studies therefore involve repeated dosing. Provided that the pharmacokinetics of a compound is linear or first-order, the same principles hold true for repeated dosing via the oral route as for repeated intravenous dosing with respect to steady-state, time to steady-state and the average concentration within a dosing interval. Repeated dose simulations are shown at two dose levels in Figure 2.35. The dosing interval is 4 hours. Case A shows that the amount eliminated is much less than the dose because there is a substantial step-up in exposure with the second dose. In Case B the amount eliminated within the 4 hour dosing interval is slightly less than the administered dose and some accumulation occurs. In Case C the amount eliminated within a dosing interval is equal to the dose administered, no further accumulation is seen, and steady-state is achieved. The

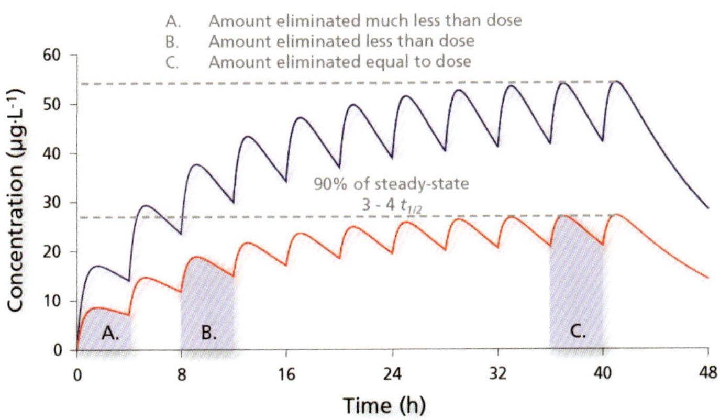

Figure 2.35
Simulated concentration-time courses after repeated oral dosing at two dose levels. It takes 3–4 $t_{1/2}$ reach 90 % of steady-state at about 24h and this time is independent of the dose provided linear or first-order pharmacokinetics prevail. 24 h divided by four gives an effective half-life of about six hours.

ratio of the grey shaded area within a dosing interval at steady-state (Case C) to the grey shaded area after the first dose (Case A) gives the accumulation factor. The accumulation factor tells you how many times higher the peak or average concentration will be at steady-state in comparison to the first dose. One may also use the ratio of peak concentration C_{max} at steady-state to C_{max} after the first dose as an index of the accumulation.

The extent of accumulation during repeated dosing can be assessed by means of Equation 2:20. The accumulation factor R_a is the extent of accumulation between first dose and steady-state. In other words, how many times higher is the concentration at steady-state is in comparison to the first dose. The accumulation factor R_a can either be obtained from the ratio of the area under the curve within a dosing interval at steady-state ($AUC_{ss-\tau}$) to the area under the curve within the first dosing interval ($AUC_{0-\tau}$), or from the ratio of maximum concentration C_{max} at steady-state to C_{max} after the first dose. This expression can be rearranged for prediction of the *effective half-life* $t_{1/2eff}$.

$$R = \frac{C_{max,ss}}{C_{max,first}} = \frac{AUC_{ss}^{\tau}}{AUC_{0}^{\tau}} = \frac{1}{1-e^{-\frac{ln2}{t_{1/2eff}}\cdot\tau}} \qquad (2:20)$$

Three to four times the *effective half-life* is the time it takes for a drug to reach steady-state independently of its multi-compartmental nature or dose size, provided linear kinetics prevail.

Figure 2.36 shows four simulated concentration-time curves of a constant-rate intravenous infusion, repeated oral dosing of an instant release (IR) formulation and repeated dosing of two extended release (ER) formulations. The disposition half-life is 10 h which means that steady-state is attained after about 40 h of constant intravenous infusion. It is also evident that steady-state is established after three IR doses, but is further delayed when the two ER formulations are given. It is obvious that with the ER formulations the time to steady-state is governed by the absorption half-life which now determines the terminal slopes. Elimination becomes absorption-rate limited. The average plasma concentration at steady-state is, however, the same for all four cases. Comparing the aforementioned administration modes it is also seen that the IR formulation not only gives rise to greater peak-trough variability, but also to transient interruptions of drug coverage when exposure falls below the therapeutically effective range – a situation which is clearly non-optimal, *e.g.*, when treating infections with bacteriostatic agents that require continuous target exposure. An ER formulation may thus be used both to avoid side effects due to high plasma peaks and periods of low or ineffective plasma concentrations (drug holidays). In this context it is further worth noting the distinction between pharmacokinetic (terminal) $t_{1/2}$ and *effective* $t_{1/2}$, with the latter representing what is relevant from a biological (therapeutic) response point-of-view.

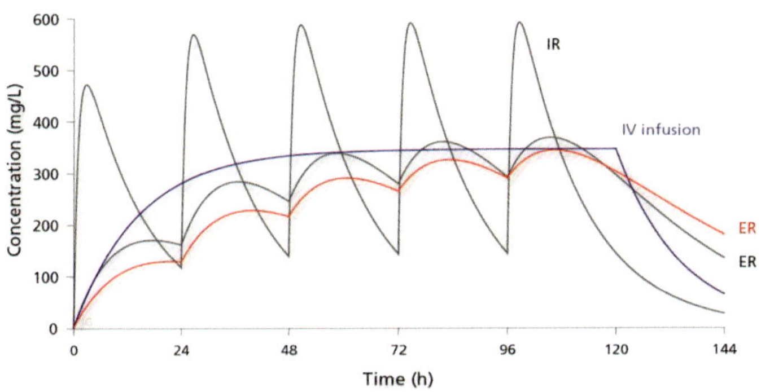

Figure 2.36
Simulated concentration-time profiles after constant rate intravenous infusion (blue curve), once a day dosing with an instant release formulation (grey curve) and after two extended release formulations (black more rapid and red slow). Note that the time to steady-state is governed by the disposition half-life (elimination half-life) for the intravenous infusion and the IR formulation. Time to steady-state is delayed with the extended release formulations because the elimination is governed by the absorption rate and its corresponding half-life.

Figure 2.37 shows three simulated concentration-time curves after repeated oral dosing. Steady-state is reached at 10–15 hours, after about three doses in this case. The standard situation is illustrated by the solid black line. If clearance is reduced by a factor of about four, the average concentration at steady-state will increase 4 times (blue dotted line). But since clearance is reduced, the half-life increases and the time to steady-state (3–4 times $t_{1/2}$) will also increase by a factor of four. Now, steady-state is reached after about 8–9 doses. The dashed red line shows an example when the volume of distribution has increased but clearance stays the same. Since the volume has increased the half-life will also increase and the time to steady-state (3–4 times $t_{1/2}$) increases. When the volume increases, the space a dose will distribute into becomes larger and hence the peak concentration will be lower. The peak-to-trough fluctuation will also be lower because the longer half-life, but the average concentration is the same as for the solid black curve.

Figure 2.37
Simulated concentration-time profiles after repeated oral dosing of compound X. The solid black line represents the standard situation, the blue dotted line when clearance is decreased by a factor of four, and the dashed red line where volume of distribution has increased. Note the increased average plasma concentration and the extended time to steady-state (blue dotted line), and the reduced peak-to-trough fluctuation and increased time to steady-state (dashed red line).

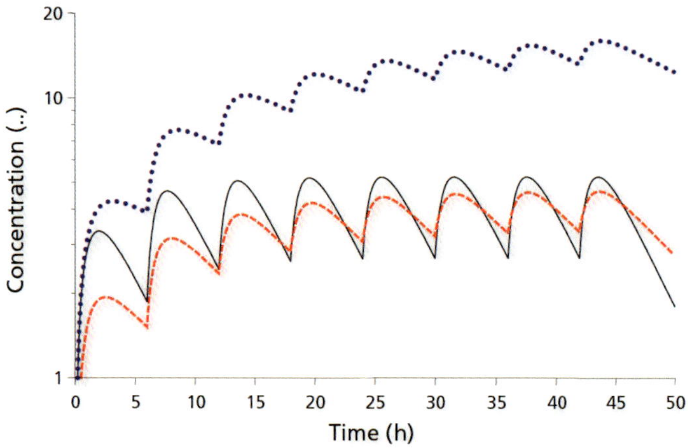

2.10 Dose-rate and unbound clearance determines free concentration

The determinants of total C and unbound C_u plasma exposure to drugs following oral dosing is shown in Figure 2.38. Note that the plasma protein binding (in terms of free fraction), protein concentration, affinity between drug and protein or the number of binding sites per protein do not affect the unbound plasma concentration after oral administration provided that there is a rapid equilibration time between drug and protein. It is the dosing rate (*DR*) and unbound clearance that are the determinants. The combination of effective plasma concentration range C_u (obtained from the *in vitro* or *in vivo* potency) and the unbound clearance Cl_u determines the *DR*. In a preclinical setting, these two parameters must be jointly optimized, not based on the free fraction f_u.

Oral dosing - Exposure

$$C = F \cdot \frac{DR}{CL} = \frac{DR}{f_u \cdot Cl_u}$$

- Dosing rate
- Unbound clearanc
- Free fraction

$$C_u = f_u \cdot C = \frac{DR}{Cl_u}$$

- Dosing rate
- Unbound clearance

Figure 2.38
Relationship between the independent variables dosing rate *DR*, free fraction f_u and unbound clearance Cl_u, and the dependent variable total plasma concentration *C* (upper row) after oral dosing. Relationship between *DR*, f_u and Cl_u, and unbound plasma concentration C_u (bottom row).

Table 2.2 summarizes the key parameters discussed in Sections 2.3 – 2.6 in terms of what they represent and how they are both measured and applied.

Table 2.2
Summary table of some pivotal absorption parameters K_a, t_{max} and F, and unbound concentration and accumulation factor

Parameter	What is it?	How is it measured?	How do you apply it?
K_a	Absorption rate constant	Determines slope of residual line. (See Figures. 2.25–2.26).	For assessment of *rate* of absorption
t_{max}	Time of maximum plasma concentration C_{max}	Read off from time axis at C_{max}. (See Figure 2.24).	For assessment of *rate* of absorption
F	Bioavailability, extent of absorption also known as fraction absorbed	$F = \frac{AUC_{po}}{AUC_{iv}} \cdot \frac{D_{iv}}{D_{po}}$	For assessment of *extent* of absorption
C_u	Unbound concentration after oral dosing of a compound primarily eliminated via metabolism	$C_u = \frac{Dose\ rate}{Cl_u}$	Optimization should be based on unbound clearance and effective unbound concentrations C_u, and not free fraction f_u
R_a	Accumulation factor	$R_a = \frac{AUC_{ss}^{\tau}}{AUC_0^{\tau}} = \frac{1}{1 - e^{-\frac{ln2}{t_{1/2eff}} \cdot \tau}}$	For assessment of how many times higher concentration is at steady-state compared to the first dose

2.11 Multi-Compartment Models

2.11.1 One- versus two-compartment models

The mono-exponential or one-compartment model is not always suitable for predicting the early time course of drug in plasma when drugs are administered rapidly and concentrations measured frequently, due to a multi-phasic decline of the concentration-time curve. The effect of adding a second or peripheral compartment to the one-compartment model is to introduce a second exponential term into the predicted time course of the plasma concentration, thereby including both a rapid initial and a slow terminal phase. This pattern is often found experimentally, and is most clearly revealed when the concentration data are plotted semi-logarithmically (Figure 2.39), because it takes time for the drug to distribute into the tissues and reach equilibrium.

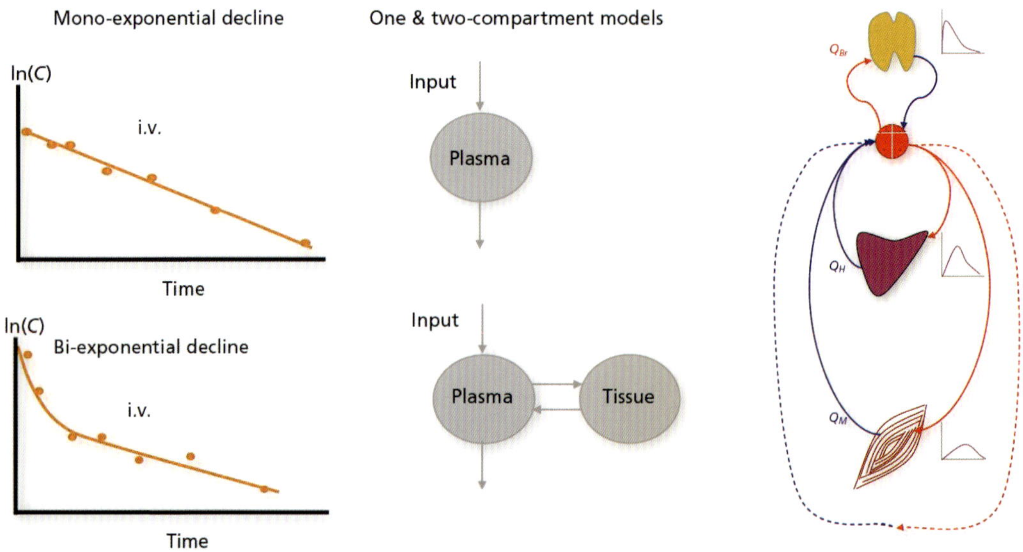

Figure 2.39
Left and Middle plots: Schematic illustration of mono-exponential decline (left, one-compartment model) and multi-exponential decline (Middle; multi-compartment model) after intravenous bolus administration.

Right: Schematic illustration of well-perfused organs (Brain and liver) and in the resting state poorly perfused tissue (muscle on-d adipose tissue). Q_{Br}, Q_H and Q_M denote brain, hepatic and muscular blood flow, respectively. The small insert concentration-time courses indicate drug profiles in well (Brain and Liver) and poorly (muscle and adipose tissue) perfused tissue.

This delay may be due to perfusion or diffusion rate limited processes, also including inward- and outward-directed transporter proteins, during drug distribution into the tissue spaces. Instantaneous equilibrium of drug between plasma and tissues is rarely observed. Most drugs display multi-compartment behaviour independently of their molecular size. Multi-compartment models are more complex than the one-compartment model because they introduce additional compartments for distribution and redistribution of the drug between a central compartment (blood and rapidly equilibrated organs) and one or several peripheral compartments (Figure 2.39). Peripheral compartments are represented by more slowly equilibrating tissues which may either be poorly perfused (adipose tissue, bone marrow), or surrounded by protective membranes (brain, foetus). In contrast to the one-compartment model, a longer time is usually required for multi-compartment compounds to achieve pseudo-distribution equilibrium between blood and tissues. It should also be borne in mind that perfusion may vary significantly depending on the actual physiological condition; *e.g.*, in a fight-or-flight situation, brain perfusion remains intact while muscles and lungs receive greater blood flow than at rest, at the expense of skin and gastrointestinal organ perfusion. Input and irreversible loss of drug is generally assumed to occur only from the central compartment. That is also a requirement for a correct assessment of the volume of distribution, since sampling occurs from the central compartment. There are, however, exceptions to this rule. The disposition of antibodies is more complex with combined peripheral elimination and slow distribution between the central and peripheral compartments. Their volume of distribution will be underestimated.

Figure 2.40 shows the behaviour of a two-compartment drug when it has been given either intravenously (left) or orally (right). In order to assess how many compartments characterize the plasma kinetics, the data are plotted on a semi-logarithmic scale. To use a stringent nomenclature, the first rapid change in the plasma concentration is called the initial phase and this is independent of whether dosing has been intravenously or oral. The last portion of the plasma concentration-time course is always called the terminal phase.

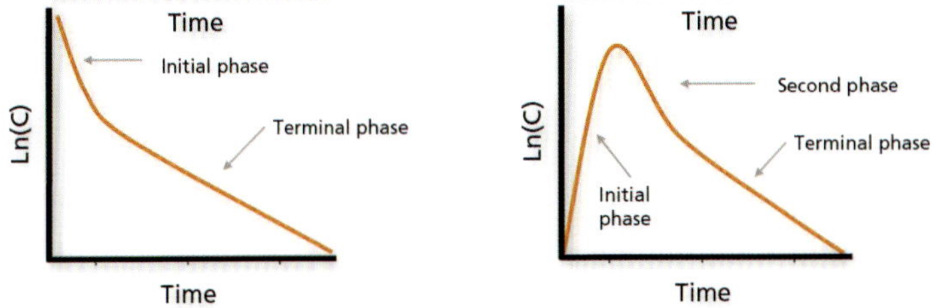

Figure 2.40
Schematic illustration of multi-exponential decline (left column, two-compartment model disposition after a rapid intravenous injection) and multi-exponential decline (right, multi-compartment model after oral administration). Upper row are plots on a linear concentration scale and bottom row show plots on a logarithmic concentration scale.

Table 2.3 shows a few examples of different tissue-to-blood partition coefficients of drugs and their resulting volumes of distribution. Of these drug examples, particularly methadone, morphine and pethidine with their extensive tissue penetration and large V_{ss} display multi-compartment behaviour in their plasma concentration-time curves.

Tissue	Methadone	Morphine	Pethidine	Theophylline	Salicylic acid
Blood	1	1	1	1	1
Brain	4.6 ± 1.2	-	17 ±0.6	0.36 ± 0.03	0.2
Renal	96 ± 19	9.5 ± 1.6	34 ± 4.6	1.21 ± 0.16	1.2
Skeletal muscle	15 ± 4.9	2.5 ± 0.5	6.3 ± 0.16	0.60 ± 0.07	0.3
Liver	71 ±15	1.2 ± 0.5	316	0.70 ± 0.08	0.5
Adipose tissue	-	-	5 ±1.2	-	-
Pulmonary	160 ± 40	-	29 ±6.2	0.71 ± 0.08	0.6
V_{ss} (L)	140 ± 910	230 ± 60	310 ± 60	35 ± 11	12 ± 2

Table 2.3
Steady state tissue-to-blood partition coefficients K_{pi} for methadone, morphine, pethidine, theophylline and salicylic acid in rats together with their respective volume of distribution at steady-state V_{ss} in man. Dash = No data available.

Drugs will display rapid kinetics in highly vascularized tissues and organs, whereas in poorly perfused tissue such as adipose tissue and bone marrow, they will display a slow turnover. In this context it should be noted that disease conditions accompanied by impaired tissue perfusion require dose adjustment to compensate for pathological/pathophysiological alterations that may otherwise significantly impact drug disposition (cf. also section 2.4.3). The volume of distribution at steady-state V_{ss} is not completely devoid of the influence of elimination, since it is a function of separate tissue or organ volume, tissue-to-blood partition coefficient K_{Pi} and also the extraction ratio E_{Ti} for eliminating organs (Figure 2.41). The organ or tissue volumes are anatomical and represent an actual weight or volume. K_{Pi} represents the physicochemical portion of V. K_{Pi} can be very large for certain compounds and tissues. K_{Pi} is also the reason V can be very large for some compounds such as quinacrine which has a value of 40,000 L. On the other hand, the lower boundary of V is the plasma volume. Thus, compounds that are highly bound to albumin, such as salicylic acid and warfarin, will also have a similar V to that of albumin (V about 0.1 L·kg^{-1}),and compounds like erythropoetin with a high molecular weight and large steric volume may be confined in plasma water only, with a V about 0.05 L·kg^{-1}, which is the lower limit of V.

Volume of distribution

$$\text{Epo}\,(0.05\text{L/kg}) \le \text{SA}(0.1\text{L/kg}) \le V_{ss} \le \text{Qcrine}(> 75L/kg)$$

$$V_{ss} = V_B + V_H \cdot K_{PH} \cdot (1 - E_H) + V_{Br} \cdot K_{PBr} + V_K \cdot K_{PK} \ldots\ldots + V_{Ad} \cdot K_{PAd}$$

Physiological/ anatomical components

Partitioning at steady-state

$$A_b = V_{ss} \cdot C \qquad V(blood\ water) \le V_{ss} \le \infty$$

Blood volume 5 L, plasma 3, total body water 40 L in man

Figure 2.41
Summary picture of factors governing the volume of distribution V_{ss}. These correspond to anatomical, physiological components like organ and tissue sizes. There are also physicochemical components such as the tissue-to-blood partition coefficients, and clearance components such as perfusion rate and intrinsic clearance. The latter two make up the extraction ratio. B denotes blood, H hepatic, Br brain, K kidney and A adipose tissue. K_{pH}, K_{pBr}, K_{pK} and K_{pAd} denote the tissue-to-blood partition coefficients of the aforementioned tissues. Erythropoetin represents the lower limit of volume equal to the plasma water 3.5 L. Salicylic acid (SA) has a volume of distribution similar to albumin of about 7 L. Quinacrine (Qcrine) exhibits a huge volume of distribution, primarily due to its extensive affinity to extravascular tissue.

2.11.2 Intravenous bolus administration

Figure 2.42 depicts the early and late time course of a drug that obeys bi-exponential decline. The net rate of transport of compound initially occurs towards the periphery $Cl_d{\cdot}C$ in parallel with irreversible loss $Cl{\cdot}C$. When pseudo-equilibrium has been established the net flux will be reversed towards the central compartment $Cl{\cdot}C_t$, which is also called redistribution, and then removed according to $Cl{\cdot}C$ rate of elimination.

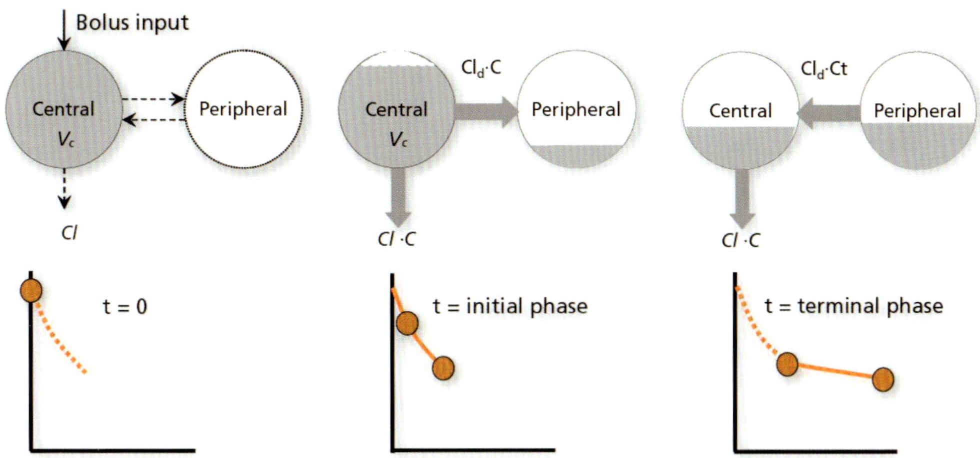

Figure 2.42
The early time course of a drug in plasma following an intravenous bolus dose (left and middle figures). The central compartment is assumed to represent plasma. The terminal phase (right figure) represents the redistribution and elimination processes.

A bi-exponential decline after an intravenous bolus dose, is described by a two-compartment model

$$C = A \cdot e^{-\alpha \cdot t} + B \cdot e^{-\beta \cdot t} \tag{2:21}$$

When data are plotted on semi-logarithmic scale the primary model parameters in Equation 2:21 are A, α, B and β, where A and B denote the two intercepts with the concentration axis and $-\alpha$ and $-\beta$ correspond to the initial and terminal slopes, respectively. Their corresponding half-lives are estimated as

$$t_{1/2\alpha} = \frac{\ln(2)}{\alpha} \tag{2:22}$$

$$t_{1/2\beta} = \frac{\ln(2)}{\beta} \tag{2:23}$$

Figure 2.43 shows the concentration-time data of a compound that displays multi-(two)-compartment characteristics following an intravenous bolus dose. A bi-exponential decline on a semi-logarithmic scale is synonymous with two-compartment behaviour. The figure demonstrates how the bi-exponential decline is dissected into its principal parts. The slope factor α is obtained by means of the method of residuals (cf. Section 2.7, Figures 2.25 and 2.26), which means that the back-extrapolated β phase is subtracted from the experimental data during the initial phase. The slope of the differences plotted on a semi-logarithmic scale corresponds to $-\alpha$. The intercept of the back-extrapolated concentration-time curve on the concentration axis C^0 is equal to the sum of the intercepts $A + B$.

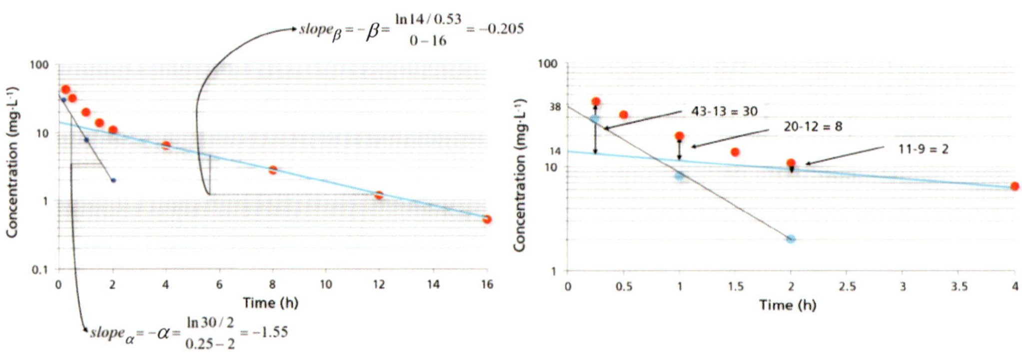

Figure 2.43
Semi-logarithmic plot of the bi-exponential decay in plasma of a drug administered as an iv bolus dose. The A parameter is about 38 mg·L^{-1} and B equals 14 mg·L^{-1}. The slope parameter α equals 1.55 and β equals 0.205, both having the unit h^{-1}. Note that the sum of the black and the blue lines corresponds to the observations. Remember that the slopes of the α and β phases are $-\alpha$ and $-\beta$, respectively.

The phase that contributes most to the total area is the elimination phase. Note that the fractional area corresponds to the fraction of the dose that is eliminated during that particular phase. In other words, the initial α phase can also be the elimination phase for some compounds (e.g., aminoglycosides), provided it contributes with the largest portion to the total area. The initial and terminal phase will be used for the respective portion of the concentration-time curve in a semi-logarithmic diagram. Unfortunately, the terminal phase is often called the elimination phase, which may not always be true. Figure 2.44 shows three different scenarios after an intravenous bolus dose. In scenario I the initial phase represents 2/3 of the total area under the curve, and therefore becomes the elimination phase. Scenario II covers a 50/50 split in fractional areas, which means that 50 % of the amount of drug is eliminated during each phase. In scenario III the terminal phase (β-phase) represents 90 % of the total area under the curve and is therefore considered the elimination phase.

Figure 2.44
Schematic illustration of how to interpret which phase (initial or terminal) corresponds to the elimination phase in a two-compartment system. Numbers within brackets represent the relative contribution of the initial and terminal phases. In plot I the initial phase contributes 2/3 of the total area and thus corresponds to the elimination phase. Note the different concentration scales in the three plots.

$C = 100 \cdot e^{-0.1 \cdot t} + 10 \cdot e^{-0.02 \cdot t}$ $\quad$ $C = 100 \cdot e^{-0.1 \cdot t} + 10 \cdot e^{-0.01 \cdot t}$ $\quad$ $C = 10 \cdot e^{-0.1 \cdot t} + 10 \cdot e^{-0.01 \cdot t}$

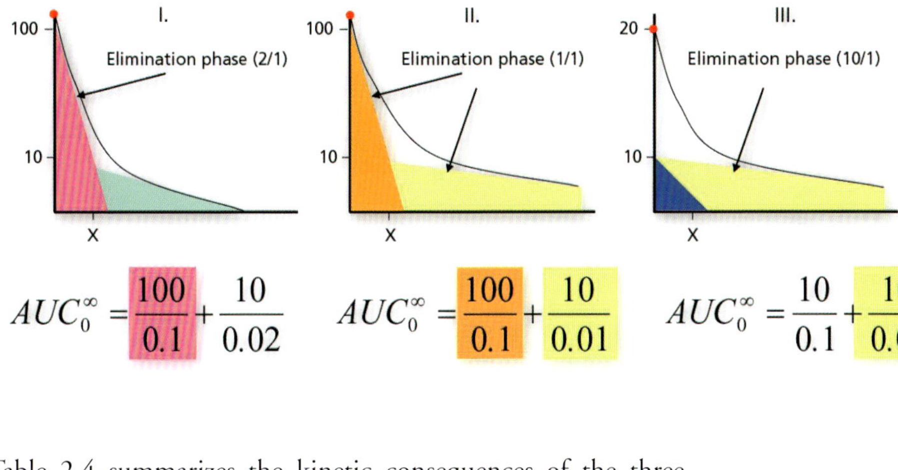

$AUC_0^\infty = \frac{100}{0.1} + \frac{10}{0.02}$ $\quad$ $AUC_0^\infty = \frac{100}{0.1} + \frac{10}{0.01}$ $\quad$ $AUC_0^\infty = \frac{10}{0.1} + \frac{10}{0.01}$

Table 2.4 summarizes the kinetic consequences of the three scenarios in Figure 2.44 with respect to mean residence time *MRT*, effective half-life $t_{effective}$ *ln(2)·MRT* and contribution of α and β phases to total *AUC*.

Table 2.4
Kinetic characteristics of models I-III in Figure 2.44 with respect to α and β-half-lives, *MRT*, effective half-life $t_{effective}$ and fractional area of the dominating phase. *MRT* denote the mean residencetime which is the average time a molecule stays in the system.

Model	$t_{1/2a}$	$t_{1/2b}$	*MRT*	$t_{effective}$	*% AUC*
I	6.9	35	23	16	a phase 67 %
II	6.9	69	55	38	50:50
III	6.9	69	92	64	b phase 91 %

Clearance is the dose divided by the total area under the curve

$$Cl = \frac{D_{iv}}{AUC} = \frac{D_{iv}}{\frac{A}{\alpha} + \frac{B}{\beta}} \tag{2:24}$$

$$D_{iv} = Cl \cdot AUC = Cl \cdot \left(\frac{A}{\alpha} + \frac{B}{\beta} \right) = Cl \cdot \frac{A}{\alpha} + Cl \cdot \frac{B}{\beta} \tag{2:25}$$

The right hand side of Equation 2:25 shows how much of the dose is eliminated during the corresponding phase. The product of $Cl{\cdot}A/\alpha$ gives the amount eliminated during the α-phase and $Cl{\cdot}B/\beta$ the amount eliminated during the β-phase. From these expressions it may be appreciated why the area (A/α or B/β) that contributes most to the total area (A/α + B/β) is called the elimination phase. The volume of the central compartment V_c, is expressed as:

$$V_c = \frac{D_{iv}}{A + B} \tag{2:26}$$

The volume of distribution at steady state V_{ss} is

$$V_{ss} = MRT \cdot Cl = \frac{AUMC}{AUC} \cdot \frac{D_{iv}}{AUC} \tag{2:27}$$

V_{ss} is the sum of the central V_c and peripheral V_t volume terms.

$$V_{ss} = V_c + V_t \tag{2:28}$$

The effective half-life $t_{1/2}(e)$

$$\begin{aligned} t_{1/2}(e) &= \ln(2) \cdot MRT \\ &= t_{1/2\alpha} \cdot \frac{AUC}{AUC_0^\infty} + t_{1/2\beta} \cdot \frac{AUC_\beta}{AUC_0^\infty} \end{aligned} \tag{2:29}$$

A scattergram of effective half-life $t_{1/2}(e)$ plotted againts terminal half-life $t_{½}$ for approximately 500 compounds is shown in Figure 2.45.

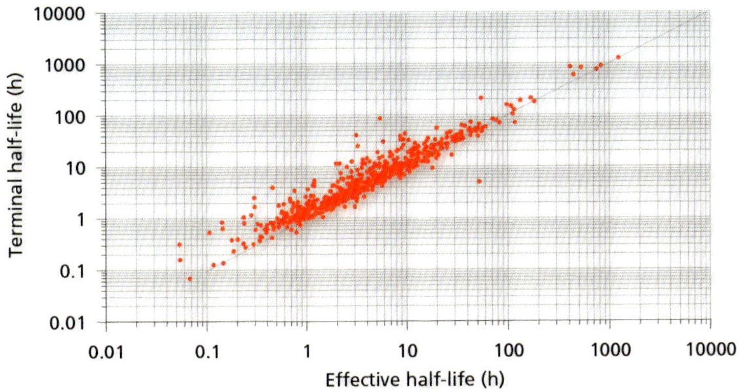

Figure 2.45
Double logarithmic plot of effective half-life versus terminal half-life of approximately 500 compounds.

"Note that most of the plotted data are above the line of unity, and thus that terminal $t_{1/2}$. From the PD perspective, though, it is the latter that is the more relevant."

2.11.3 Impact of disease on drug disposition - an example

Superoxide dismutase (SOD) has been shown to play a role in protecting major organs against ischaemia/reperfusion injury caused by oxygen free radicals. With the recent development of recombinant human superoxide dismutase r-hSOD the possibility arose of evaluating its protective effect in an animal model of a human disease. Six normal male Sprague-Dawley rats weighing 300–400 g were used as controls. In six additional rats, the renal arteries and veins were ligated on both sides, and the mid-line incision subsequently closed. A dose of r-hSOD was (20 mg·kg^{-1}) was given as a rapid 15 s intravenous infusion. Mean plasma data from untreated and nephrectomized (clamped) animals are plotted in Figure 2.46. Notice the substantial initial drop in the r-hSOD plasma concentrations in normal animals during the first 120 min as compared to clamped animals. The terminal half-lives of the two groups are similar.

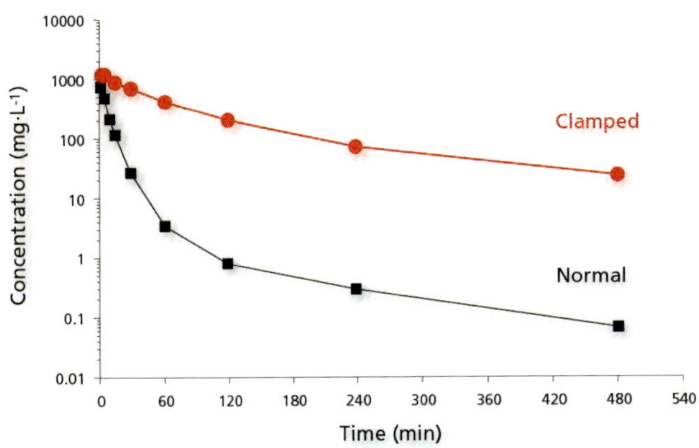

Figure 2.46
Semi-logarithmic plot of observed (symbols connected with straight lines) concentration (mg·L^{-1}) data versus time of r-hSOD following a rapid intravenous infusion of 20 mg·kg^{-1} over 15 s to normal (filled black squares) and clamped (filled red circles) rats (left).

A two-compartment disposition model is used and fitted to data obtained from normal and clamped rats. That model is represented by Equation 2.30.

$$C = A \cdot e^{-\alpha \cdot t} + B \cdot e^{-\beta \cdot t} \tag{2:30}$$

The model parameters are plasma clearance *Cl*, the volume of the central compartment V_c, intercompartmental distribution Cl_d, and the volume of the peripheral compartment V_t. The analysis is started by estimating the macro-constants *A*, α, *B* and β. The initial parameter estimates of V_c, Cl_d, V_t and *Cl* for clamped animals were derived according to the following principles. The volume of the central compartment V_c for clamped animals is estimated as

$$V_c = \frac{D_{iv}}{A+B} = \frac{20000}{1000+200} \approx 17\ mL \cdot kg^{-1} \tag{2:31}$$

Plasma clearance *Cl* is dose divided by area

$$Cl = \frac{D_{iv}}{\frac{A}{\alpha}+\frac{B}{\beta}} = \frac{D_{iv}}{AUC} = \frac{20000}{95000} \approx 0.21\ mL \cdot min^{-1} \cdot kg^{-1} \tag{2:32}$$

Volume of distribution at steady state V_{ss} is estimated as

$$V_{ss} = MRT \cdot Cl = \frac{AUMC}{AUC} \cdot \frac{D_{iv}}{AUC} = 24\ mL \cdot kg^{-1} \tag{2:33}$$

which is actually the sume of V_c and V_t

$$V_{ss} = V_c + V_t \tag{2:34}$$

which rearranged yields V_t

$$V_t = V_{ss} - V_c \approx 7\ mL \cdot kg^{-1} \tag{2:35}$$

A good initial estimate of the intercompartmental distribution parameter, which can be compared to blood flow, is for the nephrectomized animals

$$Cl_d = \alpha \cdot V_c - Cl \approx 0.1\ mL \cdot min^{-1} \cdot kg^{-1} \tag{2:36}$$

where α is the slope factor of the initial phase. The approximate initial half-life of r-hSOD in clamped animals is 38 min.

Parameter	Initial estimates	Normal rats Estimate	Normal rats CV %	Nephrectomized rats Estimate	Nephrectomized rats CV %
Cl $L \cdot min^{-1} \cdot kg^{-1}$	0.2	3	10	0.22	2
Cl_d $L \cdot min^{-1} \cdot kg^{-1}$	0.1	0.12	30	0.09	10
V_c $mL \cdot kg^{-1}$	17	31	20	16	3
V_t $mL \cdot kg^{-1}$	7	15	30	13	10

Table 2.5
Comparison of estimates and precision in the two groups of animals (normal and nephrectomized) after a rapid 15 s intravenous infusion of 20 $mg \cdot kg^{-1}$ of r-hSOD

It is clear from Table 2.5 that rhSOD has a limited distribution outside the vascular space since the volume of distribution at steady state V_{ss} is about 45 $mL \cdot kg^{-1}$ (0.05 $L \cdot kg^{-1}$ similar to plasma volume) in normal and 30 $mL \cdot kg^{-1}$ in nephrectomized animals. Normal rats displayed a typical bi-exponential decay of r-hSOD in plasma with an initial half-life of 4–7 min followed by a slower terminal half-life of about 2 h. In functionally nephrectomized rats the dominating terminal half-life was similar (2 h) to normal rats. The plasma clearance parameter was low (3 $mL \cdot min^{-1} \cdot kg^{-1}$) in normal rats, and only 4–6 % of that value in nephrectomized rats. Not only did plasma clearance decrease in nephrectomized animals but so too did the volume of the central compartment V_c. The latter decrease is explained by the fact that clamping animals also removes a substantial fraction of the blood volume perfusing the kidneys which is a major portion of V_c.

In normal rats the initial phase corresponds to the elimination phase, because the fractional area (A/α) represents 95 % of the total area under the concentration-time curve. In nephrectomized animals there is a 50/50 % area split between the initial and terminal phases. The terminal plasma half-life was about 100 min in both groups, whereas the effective half-life was 11 min in normal rats and 90 min in clamped animals. The concept of effective half-life has useful application when linear kinetics operates because it is related to the time necessary to eliminate 90 % of the drug from the body. The time to reach 90 % of steady state following constant rate infusion is primarily determined by the turnover of the amount of drug in the body, rather than the terminal decline of drug in plasma.

The removal of the kidneys increased the effective half-life from 10 min to 90 min which means that the time to steady state will differ by 9-fold between the two groups. It is often incorrectly concluded that the 'elimination half-life' is obtained from the terminal phase in both groups. It is therefore important to assess the effective half-life and not base any judgement on terminal half-lives in plasma. The effective half-life is also used for scaling across species (see further Chapter 7). The significance of distribution kinetics of r-hSOD in normal and clamped (simulating renal disease) rats has therefore been explained by interpretation of plasma concentration-time data following rapid *iv* injections.

2.11.4 Constant rate infusion

A constant rate infusion is preferred when a drug is given for the first time, has a narrow therapeutic index or when there is a desire to observe and evaluate a pharmacological effect. Both the bolus and extravascular input regimens are cumbersome since exposure to the drug cannot be controlled following its administration. The constant intravenous infusion regimen is also a better regimen with respect to safety. The pharmacokinetic data of normal and nephrectomized animals on r-hSOD which were detailed in Section 2.11.3 were used to design and predict the outcome of a constant rate infusion to the two groups of animals (Figure 2.47). The time to steady state is determined by effective half-life (10 min in normal and 90 min in nephrectomized animals) and the actual steady state concentration by clearance (0.3 and 0.02 $ml \cdot min^{-1} \cdot kg^{-1}$ in normal and nephrectomized animals, respectively).

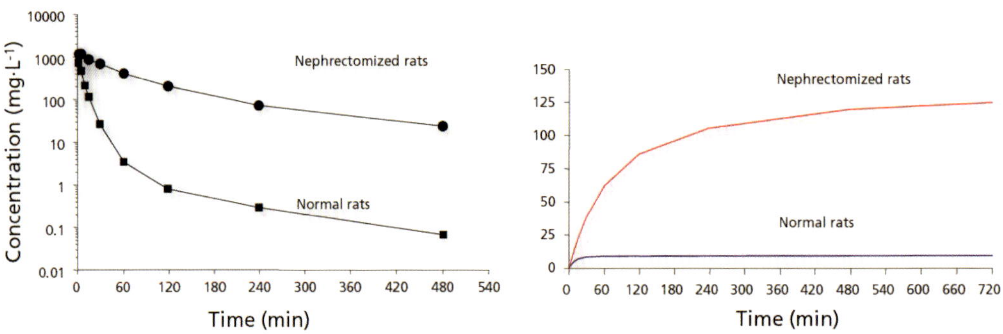

Figure 2.47
Left: Semi-logarithmic plot of observed (symbols connected with straight lines) concentration data versus time of r-hSOD following a rapid intravenous infusion of 20 $mg \cdot kg^{-1}$ over 15 s to normal (filled squares) and clamped (filled circles) rats (left).
Right: Predicted concentration-time courses in normal and nephrectomized rats after a constant rate infusion during 12 h at the same infusion rate. Note the large difference in time to 90 % of steady state and the steady-state concentration.

2.11.5 Extravascular administration

If a drug is administered extravascularly, for example orally, with an absorption process that is faster than its distribution and elimination processes, a concentration-time profile in plasma is seen, similar to that depicted in Figure 2.48. The drug concentration in plasma is zero at the time of administration and then increases rapidly as the drug enters the circulation. It reaches a maximum where the rate into the plasma compartment is equal to the rate of loss from the same compartment, and then decays in a multi-exponential fashion.

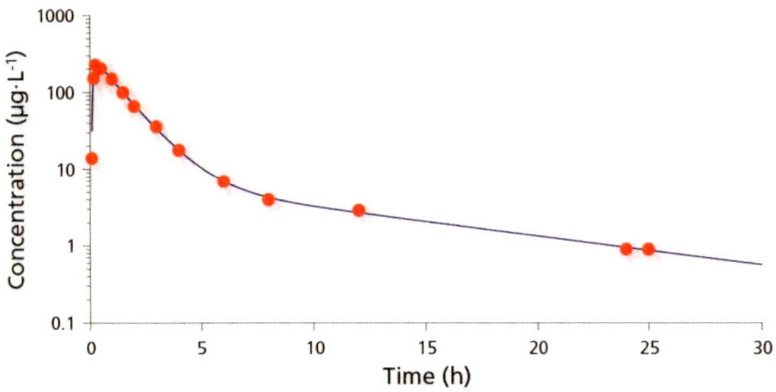

Figure 2.48
Semi-logarithmic plot of observed (symbols) and predicted (solid curve) concentration (µg·L^{-1}) data versus time of drug X after an oral dose of 20 mg. Note the rapid absorption followed by a bi-phasic decline after the peak concentration.

A bi-exponential decline after an intravenous bolus dose, is replaced by a tri-exponential model after extravascular dosing

$$C = A \cdot e^{-\alpha \cdot t} + B \cdot e^{-\beta \cdot t} - (A + B) \cdot e^{-K_a \cdot t} \tag{2:37}$$

Figure 2.49 shows the simulated repeated dose data of compound X (given as a single dose in Figure 2.48) administered every 24 h as an ordinary instant release IR dosage form (blue line) and as an extended release ER (red line) dosage form. Note how the fluctuation within a dosing interval is dramatically reduced when given as an ER formulation, and how the terminal half-life is increased with the ER formulation due to absorption rate limited elimination.

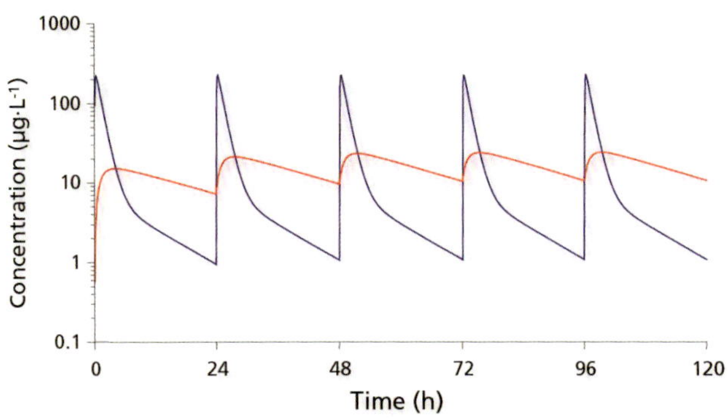

Figure 2.49
Semi-logarithmic plot of simulated concentration-time courses after repeated dosing of 20 mg every 24 h. Blue curve shows the instant release IR formulation and red curve the extended release ER formulation.

This means that the terminal half-life is governed by the absorption rate of drug. The slope of the terminal phase gives the absorption-rate constant K_a (t^{-1}). If the pronounced fluctuation with high plasma concentration peaks poses a safety problem, one may improve the safety margin by means of the ER formulation (cf. also section 2.5). The two simulations (red ER and blue IR formulations) demonstrate how to avoid rapid fluctuations with high and potentially toxic peaks and low troughs, without altering the average concentration and this AUC within a dosing interval.

2.12 Active metabolites

It is common that drugs have one or more active metabolites. The active metabolite may be part of a therapeutic strategy, for example the use of a prodrug; L-DOPA and omeprazole are good examples of this approach. L-DOPA, but not its active metabolite dopamine, readily penetrates the blood-brain-barrier (BBB) and is decarboxylated to dopamine in the brain where it then acts on its receptors to ameliorate the symptoms of Parkinson's disease. The H^+, K^+-ATPase inhibitor omeprazole is trapped and bioactivated in the parietal cells of the gastric mucosa, resulting in a very selective antacid action. The drug addict also makes use of a prodrug principle when injecting heroin, rather than its active metabolite morphine. Heroin is rapidly distributed to the brain where it is hydrolyzed to morphine but also another active metabolite. In this latter case, however, heroin, morphine, as well as the active metabolite all contribute to the effect, as compared to the L-DOPA and omeprazole examples, where the parent compound is ineffective *per se*.

A more recent interesting example of an active metabolite is provided by the non-selective muscarinic acetylcholine receptor antagonist tolterodine, which was developed for treatment of urinary incontinence. The hepatic enzyme CYP2D6, for which there is evidence of genetic polymorphism, is an important metabolic pathway for this compound. CYP2D6 metabolism of tolterodine results in the formation of the active and equipotent metabolite 5-OH-tolterodine. Consequently, bioavailability of parent tolterodine is large in poor metabolizers and there are low or negligible amounts of 5-OH-tolterodine. In contrast, plasma levels of 5-OH-tolterodine are high in rapid metabolizers and levels of the parent compound correspondingly low. Clinical efficacy of the drug is therefore maintained under both conditions.

More commonly the active metabolite may represent unwanted or unexpected aspects of drug treatment. The metabolite of an active parent compound can cause either unexpected effects or drug interactions in at least two ways. Firstly, an active metabolite may display a similar pharmacodynamic profile as the parent compound, but have a different pharmacokinetic behaviour. Examples of this include the metabolites nor-fluoxetine and 6-hydroxy-risperidone of the antidepressant fluoxetine and the antipsychotic risperidone, respectively. In both cases the active metabolite has a longer half-life in plasma than that of the parent compound which may be therapeutically useful, but may also produce unexpected interactions, due to genetic polymorphism in hepatic cytochrome P450 enzyme activity. Secondly, the active metabolite may have a different pharmacological profile from the parent compound. The 5-HT_{1A} receptor agonist buspirone is one example. Its major metabolite 1-(2-pyrimidinyl)-piperazine (1-PP) has prominent α_2 adrenoceptor blocking properties. Thus, the pharmacodynamic profile of the active metabolite becomes an important factor in interpreting the mechanism of action of this anxiolytic agent. Furthermore, potential drug interactions at the pharmacodynamic level will be different for the parent compound and its metabolite.

Figure 2.50 shows data on a 5-HT_{1A} receptor agonist, ebalzotan, given to rats via the oral and subcutaneous routes. Both the parent compound and its active metabolite were measured simultaneously in plasma. The suppression of rearing behaviour, a result of 5-HT_{1A} receptor activation, was used as a functional index of drug present in the brain since the more suppressed the rearing (the lower the value) the better the effect of the compound. The plasma concentration of parent ebalzotan and its metabolite was also measured following ebalzotan administration by both subcutaneous and oral routes. When the concentration of ebalzotan was plotted against rearing there was a distinct separation between subcutaneous and oral data. However, if the active metabolite concentration was plotted the rearing were similar when either data of the two administration routes were used (Fig. 2.47). This suggests that the active metabolite is more potent than parent compound, and that the first-pass metabolism of the parent compound to metabolite is substantial after oral dosing. This example is a good demonstration of how preclinical pharmacodynamics can provide a means in advance of clinical use of determining the potential conditions under which a drug may be subject to profound changes in its concentration-response relationship.

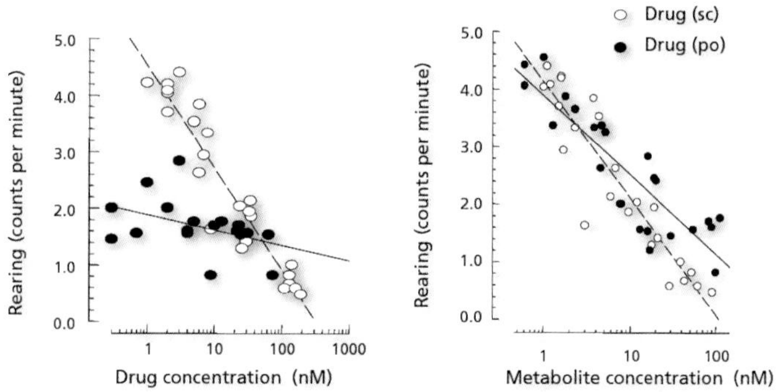

Figure 2.50
Left: Rearing plotted versus parent drug concentration after subcutaneous (open circles) and oral (filled circles) dosing of ebalzotan in rats. **Right**: Rearing effect versus metabolite concentration after subcutaneous (open circles) and oral (filled circles) dosing of ebalzotan.

2.13 How is PK and PD integrated?

This Chapter has discussed the impact of clearance and volume on the biological half-life. It also covers how clearance governs the area under the plasma concentration-time curve and the average concentration at steady-state. The (unbound) exposure (C_{max}, AUC and C_{ss}) will then drive the pharmacological response. But the biological half-life must also be put into an exposure *versus* EC_{50} perspective. Changes in half-life caused by changes in volume or clearance, have an impact on the pharmacological response when exposure is much less than the EC_{50} value, but otherwise less so. Clearance is combined with effective plasma concentrations when the human dose is estimated. Some illustrations have also been given in this Chapter, as to how diverse pharmacokinetic and physiological factors (such as routes and modes of administration, drug tissue affinity, disease states (e.g., nephrectomy), active metabolites, etc.), may impact the pharmacodynamic outcome of drug treatment, and therefore also the design of treatment protocols. Protein binding will be incorporated into the big picture in Chapter 3, where it is discussed how changes in binding affects total plasma concentrations but not the unbound concentrations. Protein binding changes affects the slope and placement of the total plasma concentration-response curve, but not the unbound concentration-response. Temporal differences between concentration and response is further discussed in Chapter 6.

Chapter 3

Principles of Plasma Protein Binding

3.1 What is this chapter about?

Drugs bind to plasma proteins to varying extents, and an important driver of the pharmacological effect is the *unbound* plasma concentration. In this Chapter the three defining parameters relating to plasma protein binding (primarily assessed experimentally *ex vivo*) are the main focus of consideration:

- the *total* (C) and *unbound* (C_u) plasma concentrations and the *free fraction* (f_u) of a drug in plasma:
  - ✓ C (*e.g.*, μmol·L^{-1}) denotes the sum of unbound drug and drug bound to plasma proteins
  - ✓ C_u (*e.g.*, μmol·L^{-1}) denotes the concentration of the drug that freely moves in the plasma water not attached to any plasma components
  - ✓ f_u (no units) denotes the ratio of unbound-to-total plasma concentrations
- what determines C_u, C and f_u and how they are interrelated
- how these factors influence the interpretation of the pharmacokinetic parameters and the pharmacological effect
- how changes in f_u affects the shape of the total concentration-response profile and how C_u should be used for the assessment of the safety margin.

Firstly, we discuss the general properties of binding and how binding affects the total concentration. We then move on to *in vitro* (closed system)-*in vivo* (open system) comparisons of binding. Then, changes in free fraction and how these changes impact pharmacokinetic and pharmacodynamic parameters are examined.

3.2 Plasma protein binding - General implications

The drug binds to blood constituents (proteins, cells and lipids), circulates as unbound (free drug) C_u or distributes across membranes into the extracellular space. These are processes that occur throughout the dosing interval. It is also generally assumed that *i*) drug unbound concentration in plasma C_u equals the unbound concentration in tissue $C_{u\mathrm{Ti}}$ at steady-state, and, ii) the unbound drug concentration in tissue equals the unbound concentration at target - also known as the *biophase*.

The unbound drug hypothesis is generally accepted. However, due to transport processes across membranes (efflux/influx), intracellular clearance (metabolism) and bulk transport (CSF and lymph flows), or (semi-)irreversible drug action (target removal), the unbound plasma and tissue concentrations may differ in spite of equilibrium. Plasma C_u is nevertheless the driving force across membranes and the pharmacological effect. For small molecules with rapid (msec-to-sec) equilibrium between unbound C_u and bound C_b drug in plasma, the bound concentration is seldom more than a passive storage pool, which varies depending on free fraction, protein concentration or compound-protein affinity.

The principal challenge with plasma protein binding relates to its variability within and across different compounds, species, strains and disease models *in vivo*. If we disregard uptake of drug into blood cells, drug in plasma is either circulating as unbound compound C_u or as compound bound C_b to plasma proteins such as *albumin* (500–700 µM) and/or *α_1-acid-glycoprotein* (9–22 µM). The equilibrium processes between unbound and bound plasma concentrations and unbound and bound tissue concentrations are depicted schematically in Figure 3.1.

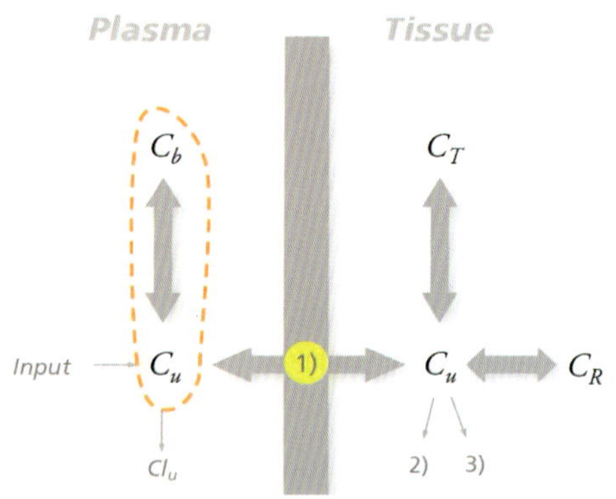

Figure 3.1
Schematic illustration of the equilibrium processes between unbound and bound drug in plasma and tissue. Deviations from a C_u plasma to C_u tissue ratio equal to unity occur when transporters (1), metabolism (2) and bulk flow (3, *e.g.*, cerebrospinal fluid flow) act on unbound tissue concentrations. In tissue, drug is unbound (C_u), nonspecifically bound (C_{Ti}) or bound specifically to target (C_R). The total plasma concentration of drug is circled by the dashed orange line.

The equilibrium between unbound and bound drug for a low molecular weight compound in plasma is determined by the unbound concentration C_u, the concentration of the binding protein $[P_T]$, the number of binding sites n per protein molecule and the affinity between drug and protein K_a. The unbound concentration C_u is determined by the (oral) dosing rate and unbound clearance Cl_u (Equation 3:1).

$$C_u = \frac{Dose\ rate}{Cl_u} \tag{3:1}$$

Consequently, to define an appropriate dose rate in man, the parameters that should be optimized during drug development in the Lead Optimization phase are the combined pharmacodynamic and pharmacokinetic properties of effective unbound concentrations C_u and unbound clearance Cl_u, respectively (Equation 3:3). Thus, it is reasonable to optimize on the product of unbound clearance and unbound effective concentrations (such as some potency measure); that is, Cl_u *and* C_u, respectively. Note that the free fraction f_u does not determine the dose rate; further expanded on below:

$$Dose = \tau \cdot Cl_u \cdot C_u \tag{3:2}$$

where τ is the dosing interval. If, for example, Cl_u is 1.5 L·min^{-1}, C_u is 200 nmol·L^{-1} (molecular weight =350 g/mol^{-1}) and the dosing interval is 24 h, the daily dose is

$$Dose(mg\ per\ day) = 1.5 \cdot (60 \cdot 24) \cdot 200 \cdot 10^{-9} \cdot 350 \approx 150\ \text{mg} \tag{3:3}$$

The total plasma concentration C may be viewed as the ratio of unbound plasma concentration-to-free fraction

$$C = \frac{C_u}{f_u} \tag{3:4}$$

where the f_u is the proportionality factor between unbound C_u and total C plasma concentrations.

Equation 3:4 is unfortunately one of the most misinterpreted and misunderstood relationships in pharmacokinetics. Therefore, when conceptualizing dependency and functionality, this equation *should not be rearranged.* The total plasma concentration C is a consequence of C_u and f_u. Experimentally, the free fraction f_u is derived by measuring the unbound C_u and total C concentrations at several concentrations, generally by means of equilibrium dialysis.

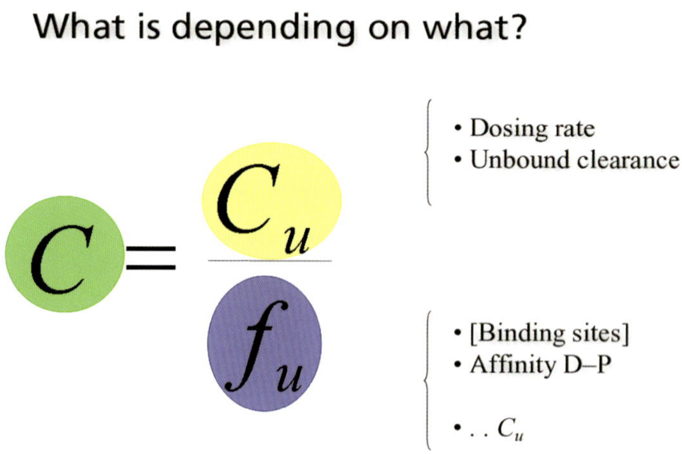

Figure 3.2
The total plasma concentration C is a function of the unbound plasma concentration C_u and the free fraction f_u. C_u is dependent on the dosing rate and unbound clearance Cl_u upon oral dosing, provided the fraction absorbed is 100 %. The free fraction f_u is dependent of the concentration of binding sites $n \cdot [P]$, affinity between drug and protein, and at high plasma concentrations also unbound concentration C_u. The n and P denote the number of binding sites per protein molecule and the concentration of protein, respectively.

The *free fraction* f_u is generally constant at pharmacological concentrations, but becomes nonlinear with respect to unbound plasma concentrations when approaching the total concentrations of binding sites on the plasma protein. There are three "capacity" regions of particular interest. At low unbound plasma concentrations binding is constant and f_u is independent of C_u. Then total plasma concentrations increase in proportion to an increase in unbound concentrations since f_u also increases slightly. At higher unbound concentrations f_u starts to increase as more binding sites are occupied. Then total drug concentration increases less than proportional to an increase in unbound drug concentration since the free fraction also increases. At high unbound concentrations, the binding proteins become saturated, f_u approaches unity, and total and unbound concentrations approach each other asymptotically. Depending on the binding plasma protein, this nonlinearity starts to occur at different plasma concentrations, often starting at only 10 % of the molar concentration of the binding protein.

For *albumin*, this is seldom an issue because *albumin* occurs in plasma at about 500–700 µM. However, for α_1*-acid glycoprotein*, a protein with a low and highly variable plasma concentration (9–22 µM in man), the risk for nonlinear binding begins already in the 1–3 µM concentration range. For many compounds f_u varies across species and strains depending on, *e.g.*, the plasma albumin concentration (Table 3.1).

Table 3.1
Data on albumin concentrations in various species and strains

Species	Strain	Mean[a] (g·L^{-1})	Mean (µmol·L^{-1})
Mouse	CD-1	30–34	455–515
Rat	Sprague-Dawley	38–44	576–667
	Wistar	29–33	439–500
Guinea Pig	Hartley	24–27	364–409
Monkey	Rhesus	43–44	652–667
Dog	Beagle	29–34	439–515
Minipig	Göttingen	29–31	439–470
Human		40–51	606–773

The *in vitro* measurement of plasma protein binding is part of the early characterization of pharmacokinetic properties of a compound. However, plasma protein binding is often limited to a crude qualitative score of low (green), intermediate (yellow) and high (red) binding. Since these are not quantitative data they cannot be used for an improved assessment of unbound concentrations, pharmacokinetic parameters, pharmacodynamic parameters (unbound potency, steepness) and safety margins. The use of quantitative *ex vivo* plasma protein binding, obtained at a minimum of two presumed concentrations (low and high), is therefore preferable as an integral part of the *in vivo* pharmacology protocol since it enables conclusions to be drawn about the impact of binding on EC_{50} and steepness of the concentration-response relationship. This becomes particularly important for exposures in the safety/tox range. Table 3.2 shows some commonly measured plasma proteins, their turnover characteristics, molecular weights and plasma and CSF concentrations, respectively.

Table 3.2
Data on properties of selected proteins in human plasma and cerebrospinal fluid

Protein	Half-life (days)	Molecular weight	Plasma ($g{\cdot}L^{-1}$)	CSF ($g{\cdot}L^{-1}$)	Comment
Albumin	15–19	66,300	35–52	0.155	General transport protein; maintains plasma osmotic pressure
α_1-acid glycoprotein	5	40,000	0.5–1.2	0.0035	Binds cationic drugs and hormones
α_1-fetoprotein		69,000			Principal foetal protein; albumin analogue
Haptoglobin	2	85,000–840,000	0.3–2.0	0.00224	Binds hemoglobin
Transferrin	7	79,600	2.0–3.6	0.014	Transports iron
Plasminogen			0.70	0.00025	Precursor of plasmin, fibrinolysis factor
Fibrinogen			2.6	0.0006	Blood coagulation
IgG	24	144,000–150,000	7–16		Antibody
IgA	6	160,000	0.7–4.0		Antibody
IgM	5	900,000	0.4–2.3		Antibody

It is not necessarily a given that the free fraction of the same compound is similar across different species or even within the same species. A collection of free fractions are plotted versus the unbound plasma concentration for different species is shown in Figure 3.3. Note the seven-fold difference in free fraction at a fixed unbound concentration of 0.1 nM between spieses, and the concentration dependent change in f_u in a single species with increasing unbound plasma concentrations.

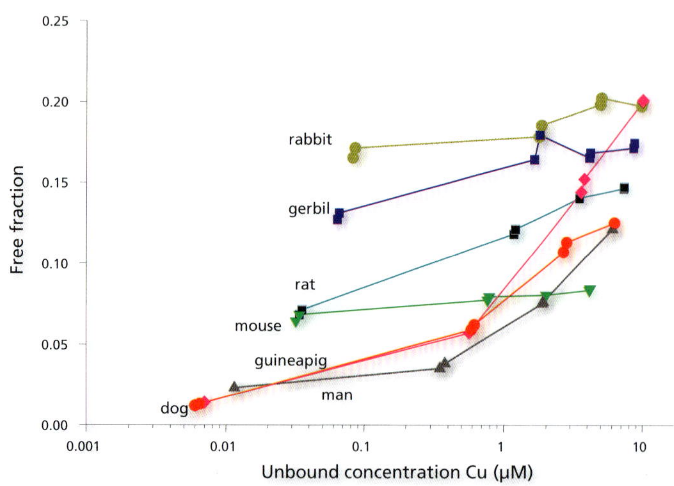

Figure 3.3
Free fraction versus unbound plasma concentration of compound X in the mouse, gerbil, rat, guineapig, rabbit, dog and man. Notice the large between species differences and the concentration-dependent change in free fraction.

3.3 *In vitro* (closed) versus *in vivo* (open) systems

An example to illustrate the principle of equilibrium and the challenges of extrapolation from *in vitro* (closed system) to *in vivo* (open system) is shown in Figures 3.4 and 3.5. Displacement from plasma binding sites has previously often been misunderstood. Alone it will only lead to a temporary and modest increase in the unbound plasma concentration and related pharmacological effect. Clinically significant interactions earlier ascribed to displacement (*e.g.*, warfarin and phenylbutazone) may in fact be explained by inhibition of drug metabolism instead. Although a number of articles have appeared that question the clinical importance of changes in plasma protein binding, both preclinical and clinical scientists involved in the drug development process continue to be concerned about the relevance of drug-drug interactions and disease-drug interactions that change the unbound plasma concentration.

There is a common misconception, often drawn from *in vitro* binding studies (*closed* systems), that the larger the plasma protein binding, the smaller is the pharmacological response *in vivo*. Now consider the schematic illustration in Figure 3.4 of three different *in vitro* binding systems.

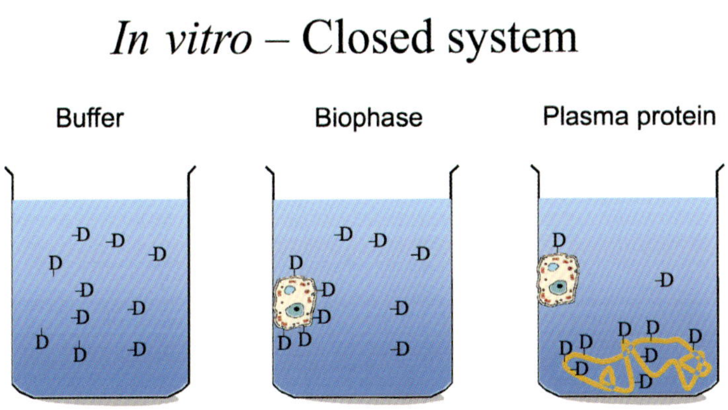

Figure 3.4
Left: Schematic illustration of three *in vitro* binding systems with the same amount of drug D (10) added to give a total concentration of 10 units. The left hand beaker shows drug added to the buffer, with all of the drug being unbound since there are no binding components.
Middle: This beaker shows the *in vitro* pharmacological model – the cell system. The cell system exhibits a good response since five drug molecules are bound to each cell. The stoichiometric equilibrium between bound and free drug is 1:1.
Right: This beaker shows a cell system, with plasma protein added to the solution. Still the stoichiometric equilibrium between free drug and drug bound to the cell system is 1:1, although much less of the total drug binds to the cell system. The cell system exhibits a lower response due to fewer drug molecules being attached to each cell. Much less drug is circulating in free form since equilibrium is shifted towards the plasma protein. Since the plasma protein binding is very high, very litte drug is now accessible to the cell system in this closed *in vitro* binding system.

The conclusion to be drawn from the *closed in vitro* binding system is that adding a competing binding protein to an *in vitro* cell system will probably reduce the number of drug molecules that are available for cellular binding. However, this is a natural consequence of a *closed* system with no input and elimination.

Now instead consider the *open in vivo* binding system illustrated in Figure 3.5. *'Open'* means that drug is added to the system (*e.g.*, animal) at a constant dosing rate and removed with an unbound clearance Cl_u. In this example, the ratio of the two gives a free drug concentration C_u of 5 units at steady state, shown in the left hand figure. The ratio of dosing rate-to-Cl_u will not change between the three scenarios.

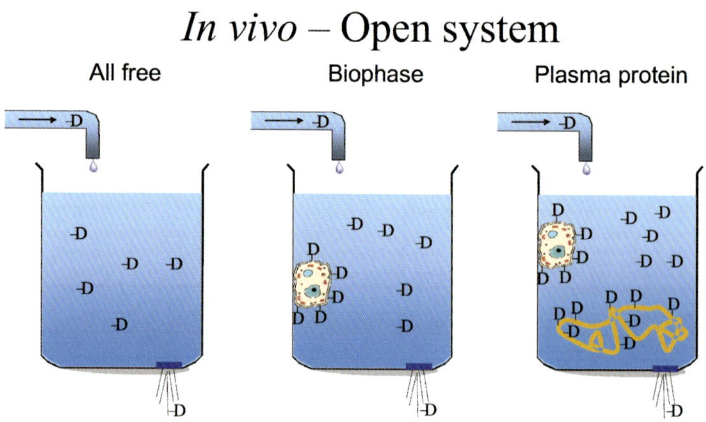

Figure 3.5
Left: Schematic illustration of three *in vivo* binding systems with the same dosing rate and unbound clearance of drug D in order to give a free concentration of 5 units in all three situations. The left hand beaker shows a system with only input and removal, resulting in a free concentration of 5 units at steady state. The total concentration is the same since there is nothing for the drug to bind to.
Middle: This beaker shows the *in vivo* pharmacological model where we also have binding to receptors at the cell surface. The cell system exhibits a good response since five drug molecules are bound to each cell. The stochiometric equilibrium between bound and free drug is 1:1. Since neither the input rate nor the elimination Cl_u are changed, the free concentration is still 5 units.
Right: This beaker shows a cell system to which plasma proteins have been added. Still five drug molecules are bound to the receptors and five units are free at steady state. However, we now have a substantial inert depot of drug also bound to plasma proteins. The cell system still exhibits the same response as in the middle figure since input rate and Cl_u are unchanged and C_u is therefore unchanged. However, the total concentration of drug has dramatically increased because of drug that is now also bound to the plasma proteins.

3.4 Impact of free fraction on PK and PD properties

This Section discusses how changes in unbound clearance and free fraction impacts total clearance and total plasma concentrations. A change in unbound clearance affects both unbound and total concentrations. Since the unbound concentration is changed, it may be anticipated that also the pharmacological response may change. Changes in the free fraction only changes total concentration, but the unbound concentration remains the same. Consequently, the pharmacological response will be unaffected. Later in this Section we will show how unbound clearance correlates with free fraction as it increases towards unity, whereas the total clearance is unaffected and falls as a horizontal band when plotted against the corresponding free fraction. Remember though that total clearance is the product of free fraction and unbound clearance. Highly plasma protein bound compounds (low f_u) have a clear tendency to also have a high unbound clearance, because they are more lipophilic compounds (Figure 3.7) it will be shown how free fraction increases (as for more hydrophilic compounds), unbound clearance decreases, and total clearance remains constant (Equation 3:5).

$$\leftrightarrow Cl = \uparrow f_u \cdot Cl_u \downarrow \tag{3:5}$$

The unbound plasma concentration C_u is governed by the oral dose rate and unbound plasma clearance Cl_u.

$$C_u = \frac{Dose_{po}\ rate}{Cl_u} \tag{3:6}$$

Since neither unbound clearance Cl_u (Equation 3.6) nor unbound plasma concentration C_u (Figure 3.2) are dependent on free fraction f_u, the oral dose $Dose_{po}$ will not be affected by changes in f_u.

$$Dose_{po}\ rate = C_u \cdot Cl_u \tag{3:7}$$

Figure 3.6 shows a simulation of unbound and total plasma concentrations in two individuals with different free fractions f_u. The dosing rate and unbound clearance is the same of the drug in both individuals giving the same unbound concentration-time course (dotted red and dashed blue lines), and consequently the same pharmacological effect. The free fraction of the drug is lower in subject B as compared to subject A, resulting in a higher total concentration of drug in subject B. The difference in free fractions may be due to different plasma protein concentrations. Since the unbound concentration is the same in both individuals the pharmacological response is expected to be the same provided same target affinity (EC_{50}) and efficacy of drug in both subjects. The two subjects then have similar unbound concentration-response relationships. Their total concentration-response relationships differ with respect to EC_{50} (placement) and potentially the slope. The total concentrations, rather than the unbound concentrations, will confound the concentration-response relationship, simply due to differences in each individual's capacity to passively store drug at the binding protein.

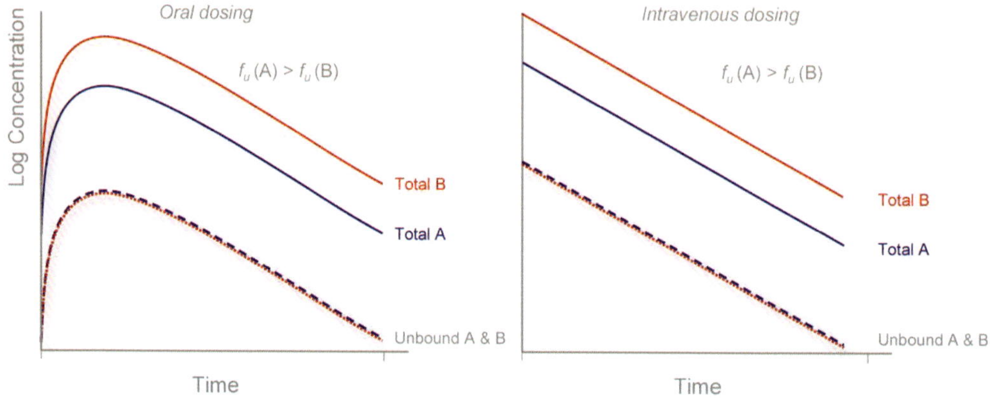

Figure 3.6
Left: Semi-logarithmic plot of unbound concentrations (dotted red and dashed blue lines) of a drug in subjects A and B after oral dosing. Since the free fraction of drug in subject A is greater than in subject B, the total plasma concentration of drug in subject B will be higher than in A. **Right**: Semi-logarithmic plot of unbound concentrations (dotted red and dashed blue lines) of a drug in subjects A and B after rapid intravenous injection. Differences in free fractions may be due to different affinities between drug and protein or different binding protein concentrations (See Table 3.1). Subjects A and B can also represent different strains like Sprague-Dawley (620 μM of albumin) and Wistar (470 μM of albumin) rats. The doses are matched with unbound clearance so that the unbound concentrations are the same.

Table 3.3 Examples of highly bound compounds, percent bound, therapeutic range and dose

Compound	Bound (%)	Therapeutic range	Dose	Unbound range
Alfentanil	92	100–340 μg/L		8–27 μg/L
Amiodarone	99.98	1–2.5 mg/L		0.2–0.5 μg/L
Amitriptyline	95	60–220 μg/L	10–50 mg	3–11 μg/L
Chlordiazepoxide	>96	>0.7 mg/L		>28 μg/L
Chlorpromazine	95–98	30–350 μg/L	25–100 mg	1–12 μg/L
Clozapin	>95		25–100 mg	
Diazepam	>98	300–400 μg/L	2–5 mg	6–8 μg/L
Digitoxin	97	>10 μg/L	0.1 mg	>0.3 μg/L
Doxazosin	>98	2.2 mmHg/μg/L		44 mmHg/ng/L
Felodipine	>99	8 nM	5–20	0.08 nM
Furosemide	>98		500 mg	
Ibuprofen	>99	10 μg/mL		0.1 μg/mL
Naproxen	>99	>50μg/mL	250 mg	>0.5 μg/mL
Nifedipine	96	47 μg/mL		2 μg/mL
Nimodipine	98			
Omeprazole	95		20–40 mg	
Phenylbutazon	>96	50–150 μg/mL		2–6 μg/mL
Warfarin	99	2.2 μg/mL	2.5 mg	20 μg/L

Table 3.3 contains information about a series of highly plasma protein bound compounds, their effective plasma concentration range and daily dose. Note that the doses are not dramatically high for these compounds. Some of the compounds (furosemide, naproxen) have daily doses above 500 mg, but that is partly explained by their low potencies and high unbound clearances, not because of a high plasma protein binding.

A compilation of clearance and volume terms of almost 500 compounds shown in Figures 3.7 and 3.8 reveals interesting patterns. Red dots show unbound clearances (Figure 3.7 left) and unbound volumes (Figure 3.7, right) and blue dots their corresponding total values plotted versus their free fractions in plasma. Note that the free fractions f_u cover three orders of magnitude. Within that range there is a tendency of a rise in Cl_u and V_u with a decrease in free fraction in spite of a substantial scatter in data. Lipophilic compounds often have a lower free fraction (higher binding) and larger Cl_u and V_u values. Note though that it is *not* implied that Cl_u and V_u are *determined* by f_u, simply that they have something in common, such as certain physicochemical properties (*e.g.*, lipophilicity).

Figure 3.7
Left: Double logarithmic scattergram of unbound clearance (red dots) and total clearance (blue dots) against free fraction of approximately 500 compounds. Note the declining central tendency of unbound clearance, and horizontal band for total clearance as the free fraction increases. The scatter in clearance across compounds at a fixed free fraction is substantial due to different chemical and pharmacological classes.
Right: Double logarithmic scattergram of unbound volume of distribution (red circles) and total volume (blue squares) against free fraction of approximately 500 compounds. Note the declining central tendency of unbound volume as compared to horizontal band for total volume when the free fraction increases. NB: These plots are not aimed to state that unbound clearance or unbound volume of distribution are dependent on the free fraction.

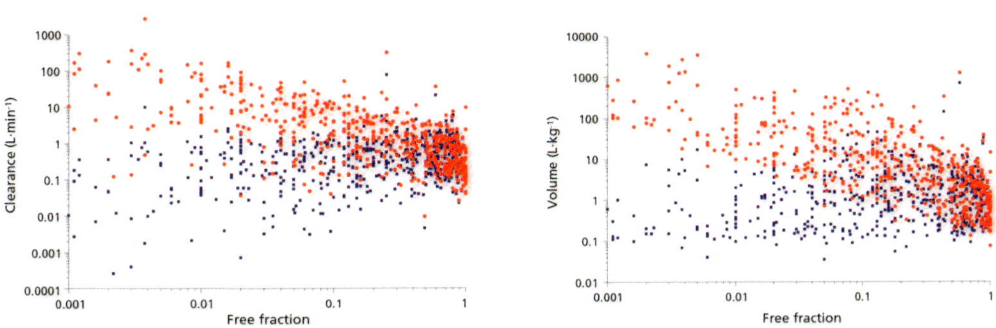

When the total (blue) and unbound (red) clearance and volume of distribution are plotted against each other, a weak correlation emerges between unbound parameter values (Figure 3.8, left). Nevertheless, the scatter is substantial and no firm conclusions should be drawn from these data alone. Figure 3.8 (right) shows a scatter-gram of effective half-life ($\ln 2 \cdot V_{ss}/Cl$) plotted against free fraction of the same compounds. There is no systematic trend in the effective half-life *in spite* of three orders of magnitude change in the free fraction.

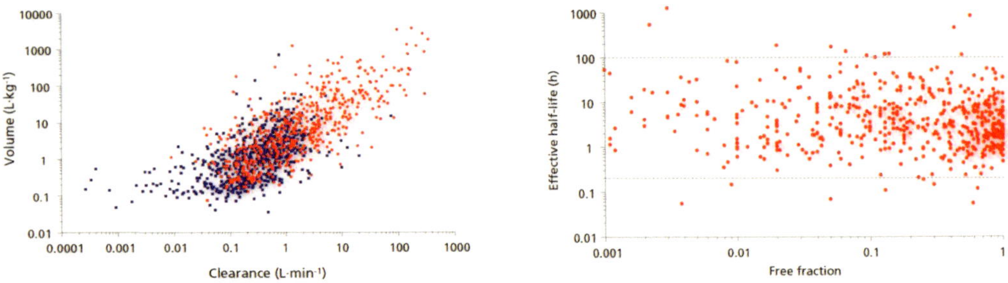

Figure 3.8
Left: Double logarithmic scattergram of total (blue) and unbound (red) clearance (left) versus total (blue) and unbound (red) volume of distribution. A weak correlation is seen between unbound parameters, although there is substantial variability. **Right**: Double logarithmic scatter plot of the effective half-life plotted versus free fraction of approximately 500 compounds. Note that the conglomerate parameter half-life is seemingly independent of the free fraction. There is no obvious tendency that half-life should change with a change in free fraction.

Often, potency measures like EC_{20}, EC_{50} or EC_{80} are derived from one species (*e.g.*, rat or gerbil) while safety measures like QT_{10} are obtained from others (*e.g.*, guinea pig or dog). As these species may exhibit different protein binding properties, it is necessary to make the comparisons based on a protein binding independent scale (*e.g.*, unbound concentrations such as EC_{u20} and QT_{u10}). For example, the same unbound concentration (*e.g.*, 100 nM) of test compound *X* for the rat (free fraction f_u 50 %), gerbil (f_u 14 %), guinea pig (f_u 18 %), dog (f_u 3 %) and human (f_u 3.7 %) will give a total plasma concentration of 0.8, 0.7, 0.5, 3.3, and 2.7 µM, respectively (Figure 3.9). Consequently comparisons across species are difficult unless based on unbound concentrations.

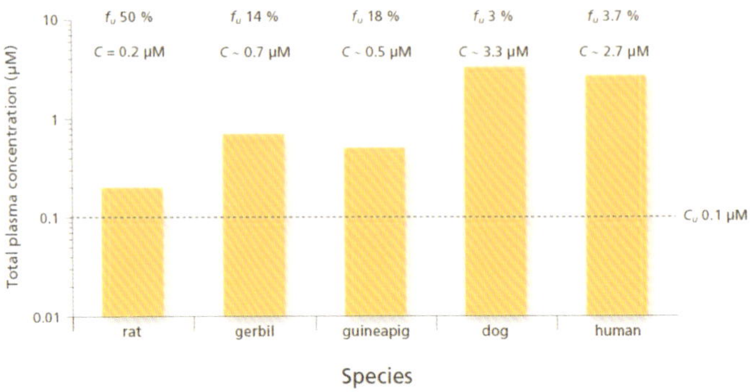

Figure 3.9
Logarithmic plot of the total plasma concentration in rat, gerbil, guinea-pig, dog and human plasma at an unbound plasma concentration of 0.1 µM. Note the 10-15 fold range in total concentration between species at this level of unbound drug.

The estimated unbound concentration giving a 20 % change from baseline of the primary effect (C_{u20}) may be compared with unbound concentrations giving a 10 % change from baseline of the secondary, safety effect (*e.g.*, C_{u10} for the *QT*-response). This enables a comparison between an efficacy and a safety measure. Figure 3.10 shows how the apparent safety margin shrinks substantially when a species-independent unbound concentration scale is used for test compound. Of course, the opposite could also occur, if the free fraction of drug in the primary model species was less than the free fraction in the safety species.

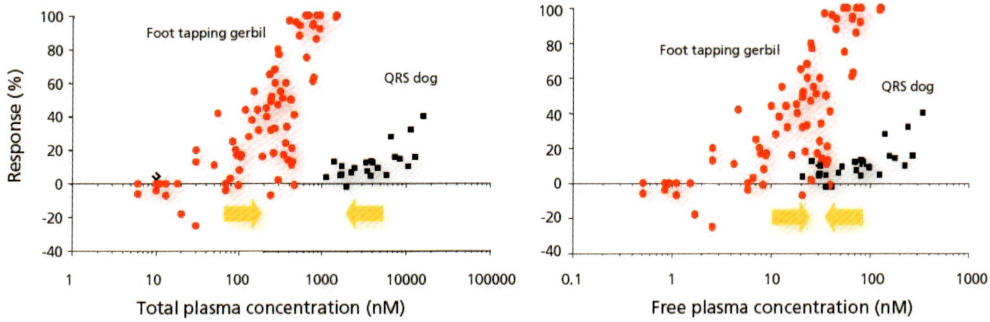

Figure 3.10
Comparison of plasma concentrations of compound X vs. effects in efficacy and safety biomarker models, represented by foot tapping in gerbils and cardiovascular (QRS) effects in the dog, respectively. The free fraction f_u is greater in gerbils than in dogs. Considering this difference between species, and plotting responses against unbound (free; right graph) instead of total (left graph) plasma concentration levels results in a decreased safety margin estimate, as indicated schematically by horizontal yellow arrows.

Figure 3.11 demonstrates how a steep total plasma concentration-response relationship becomes shallower when the unbound concentration-response relationship is used instead. The latter is the preferable measure. The reason for the differences in steepness of the two curves is a nonlinear saturable plasma protein binding. The total plasma concentration increases because the unbound plasma concentration does so. However, the free fraction increases simultaneously, which means that for the same change in response the unbound plasma concentration has to change disproportionately more than the total concentration.

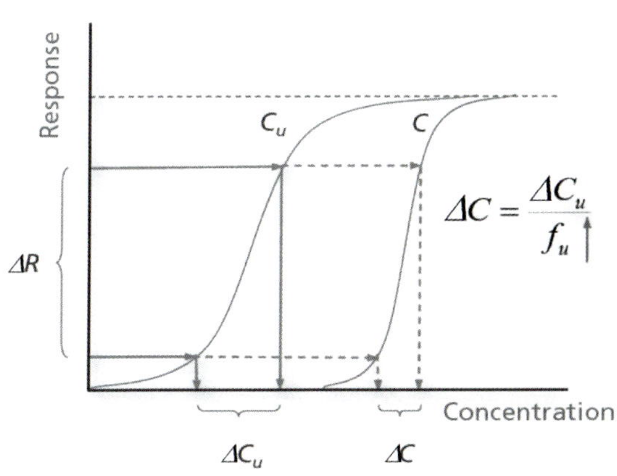

Figure 3.11
Plot of total C (right), and unbound C_u (left) concentration-response relationship. Note the greater steepness of the total concentration-response curve than of the unbound concentration-response curve. A larger change in unbound than in total concentration is required to obtain a defined increase in response DR. This is explained by the inset formula which also indicates that the free fraction changes as unbound concentration changes. The larger change in C_u is partly counteracted also by a change in the free fraction f_u, and therefore the total concentration change is smaller.

The aspects of concentration-dependent plasma protein binding and intra-assay variability in plasma protein binding are summarized in Figure 3.12 for test compound *X*. The pooled plasma for *in vitro* measurement is compared to the concentration-dependent *ex vivo* measurements obtained from the actual *in vivo* experiment.

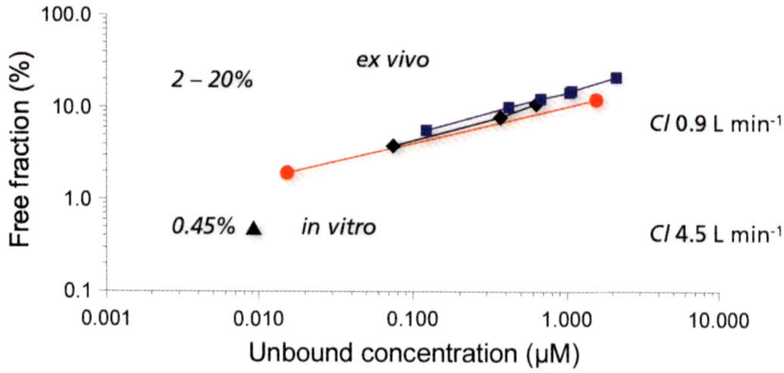

Figure 3.12
Free fraction (%) *versus* unbound concentration (blue squares dog 1, black diamonds dog 2, red circles dog 3) *ex vivo*. Also included is the *in vitro* free fraction (0.45 %) at a total plasma concentration of 2μM corresponding to an unbound concentration <0.01μM. All data from test compound X.

Using a constant *in vitro* value of free fraction (0.45 %) in Figure 3.12 will not only result in an erroneous predicted human clearance value (>4.5 L·min^{-1} in contrast to 0.9 L·min^{-1} when a concentration-dependent f_u of 2–20 % is used) but will also bias estimates of potency (EC_{u50}) and the safety margin. In contrast, a high free fraction of about 50 % in rat, guinea pig, and man further suggests species-dependent differences, and the necessity for this to be factored in when assessing kinetics, efficacy and safety. The consequences of linear *versus* nonlinear plasma protein binding on the concentration-response relationship is shown in Figure 3.13. Note the steep concentration-response relationships for total (black) and linear (blue) unbound concentrations, in comparison to the unbound concentration-response relationship (red) in the case of a nonlinear free fraction. The two examples (Figures 3.7, 3.8 and 3.13) demonstrate that plasma protein binding has an effect on both pharmacokinetics and pharmacodynamics when assessment is based on either total or assumedly constant binding. Subject variation coupled with a steep concentration-response relationship may in some cases substantially limit the practical use of a compound, particularly when the response is a risk factor (adverse event). Thus, as also evident from Figure 3.13, using the standard *in vitro*-generated linear f_u value to estimate unbound concentrations across the exposure range may overestimate potency to elicit primary efficacy *vs.* secondary, safety or toxic responses – in either case resulting in inaccurate safety margin predictions. This is particularly important as frequently different species are utilised to generate the primary *vs.* secondary/safety/toxic readouts, and the degree of free fraction linearity may well vary between species and across exposure ranges applied in different models.

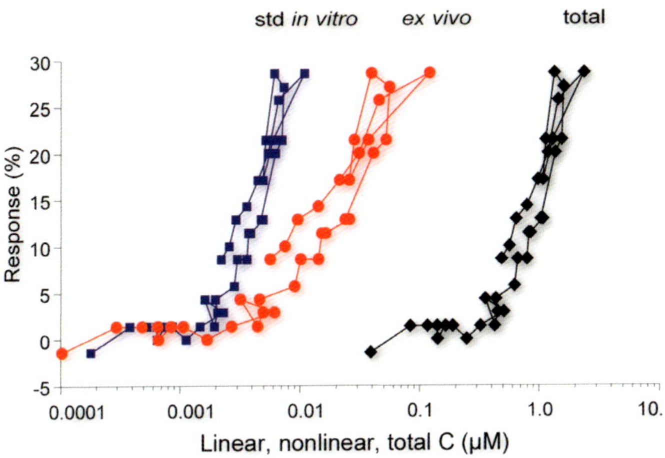

Figure 3.13
Plot of total (black), and unbound (using a constant free fraction, blue, and nonlinear free fraction, red) concentration-response relationship of test compound Y. Note that the steepness of the red curve is less than the other curves. Std *in vitro* means that pooled plasma is used for in vitro measurement. In this case a constant free fraction of 0.45 % was used to convert total (black) concentrations to unbound concentrations (blue). *Ex vivo* denote protein binding measurements done on actual plasma samples obtained from the *in vivo* experiment, in which free fraction increased with increased total concentrations. The blue, red and black data in this plot were taken from one of the dogs in Figure 3.12.

An additional example of switching from unbound C_u to total C_u/f_u when assessing the free fraction, is given in Figure 3.14. A straight linear relationship (f_u *versus* C_u, left) disappears when total plasma concentration *versus* free fraction is plotted (right). Remember that the total plasma concentration is actually the ratio of C_u-to-f_u. The variability in total plasma concentration C caused by the individual variabilities in both C_u and f_u, will be amplified when plotted as a ratio.

Figure 3.14
Left: Free (unbound) plasma concentration plotted against the free fraction. Note the distinctive straight line with very little scatter in data.
Right: Total plasma concentration plotted against the free fraction, yielding an unclear relationship between the two and substantial scatter in data.

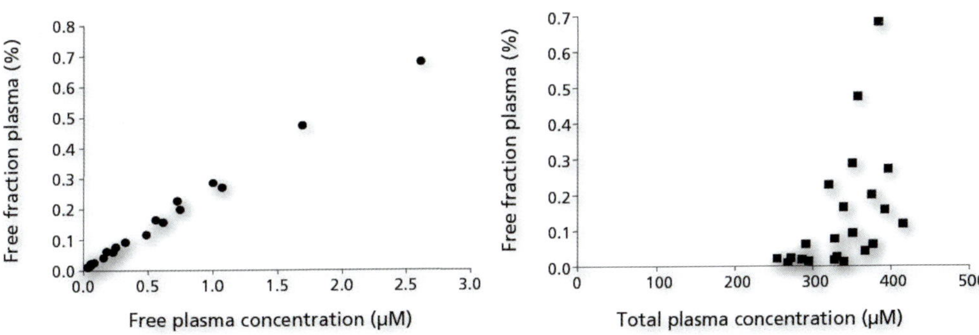

Confounding factors known to affect plasma protein binding are species and strain differences, certain disease states and conditions, concomitant use of other drugs, anaesthetics, and surgical stress. Therefore, *ex vivo* measurement of plasma protein binding in the actual pharmacological or pharmacokinetic *in vivo* models should be made, in order to avoid quantitative (concentration-dependencies) and qualitative (*e.g.*, the commercial source of proteins, presence of proteins in *in vitro* assays, etc.) differences such as those shown in Figures 3.12 and 3.13.

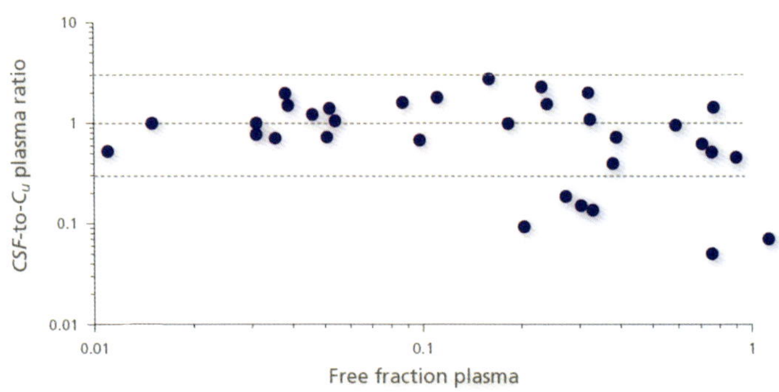

Figure 3.15
Scattergram of CSF-to-C_u plasma ratio *versus* free fraction of drug in plasma. No obvious correlation is seen (Maurer *et al* [2005]).

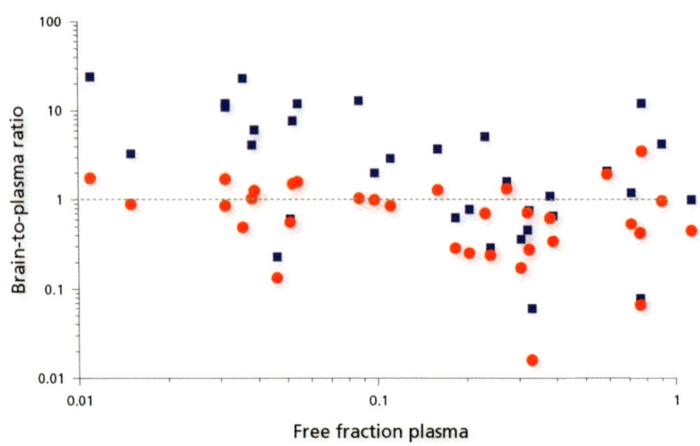

Figure 3.16
Scattergram of unbound (red) and total (blue) brain-to-plasma concentration ratios plotted *versus* free fraction (data from Maurer *et al* [2005]).

Literature data on brain penetration of 33 CNS drugs revealed that cerebrospinal fluid is a relatively poor predictor of brain uptake, correlates poorly with free fraction in plasma and contains a considerable amount of variability (Figure 3.15). The unbound brain concentration $C_{u,\ brain}$-to-the unbound plasma concentration $C_{u,\ plasma}$ ratio also displays a large amount of scatter when plotted against the free fraction in plasma (Figure 3.16). This ratio seems to be centered around unity for the entire range of free fractions. In light of these data one should be cautious when selecting or deselecting compounds based on either CSF or even unbound brain concentrations unless coupled to a pharmacological response. Partitioning to brain of these drugs measured as either total or unbound concentrations, poorly reflect their therapeutic value.

3.5 How is information about PPB integrated with PD?

Based on the examples outlined in this Chapter it may be concluded that changes in plasma protein binding may affect pharmacokinetic parameters, but will usually not change the clinical outcome in patients. A consequence of this is that no dose adjustment will be necessary. The only exception to this would be in rare cases for drugs with a narrow therapeutic window and a very rapid kinetic/dynamic equilibration time, that are given intravenously and have a high extraction ratio (or, even more rarely, with such properties and given orally). We are only aware of one (intravenously administered) compound fitting this description, and that is bupivacaine. With this as an exception, one may conclude that the unbound drug concentration is dependent on the protein concentration in a *closed system (in vitro)*, while the unbound drug concentration is independent of the plasma protein concentration or tissue binding in an *open system (in vivo)*, provided a rapid equilibrium between free and bound. When a highly protein-bound compound X is displaced from the binding protein by another compound Y, a transient sharp increase in the unbound concentration of X may occur, leading to adverse effects. A typical commonly cited drug interaction is between warfarin and phenylbutazone. Because both compounds display high plasma protein binding (warfarin f_u 0.01 and phenylbutazone f_u 0.03), their kinetic interaction has been attributed to an interaction at the level of their plasma protein binding. However, recent data provide evidence that the metabolism of both compounds is mediated via the same isozyme, *CYP2C9*, and clinically significant inhibition that may occur with coadministration. Individualization of dose is therefore essential for *CYP2C9* substrates that have a narrow therapeutic index.

One of the take home messages from this Chapter is that changes in the binding protein concentration will change the free fraction. The free fraction should ideally be viewed as a nonlinear proportionality constant between total and unbound concentrations. The free fraction is determined by the affinity constant K_a between protein and drug, and the total concentration of binding sites $n \cdot [PT]$, as well as by the unbound concentration C_u when the molar concentration of the drug becomes greater than 10 % of the molar concentration of the binding protein. The total concentration (dependent variable) is a function of unbound concentration (independent variable) and the free fraction. Ideally, drug selection should be based upon a low unbound clearance and a high potency (low effective unbound concentration) or a combination of these two properties. This should also guide the structural modification of new compounds. It is highly recommendable to exercise caution when extrapolating from a closed (*in vitro, e.g.,* overexpression of target) system to an open (*in vivo*), unless the determinants of unbound concentration (Cl_u and dosing rate) are taken into account. Table 3.3 summarizes the key parameters discussed in this Chapter in terms of what they represent, how they are measured, and their particular impact.

Table 3.3
Summary table of the characteristics of C_u, f_u and C

Parameter	What is it?	How is it measured?	Comments
C_u	Unbound plasma concentration (units e.g., µM)	Equilibrium dialysis, ultra filtration	For assessment of impact of disease, concomitant drug use and for safety margin. C_u drives the pharmacodynamics and drives *in vivo* efficacy
f_u	Free fraction (no units)	Experimentally as the ratio of unbound C_u-to-total C plasma concentrations $f_u = \frac{C_u}{C}$	Impacts the total concentration-response relationship, placement and slope
K_{Pu}	Tissue-to-Unbound plasma concentration ratio	Experimentally as the ratio of unbound C_u-to-total C plasma concentrations	A measure of tissue penetration independent of plasma protein binding
C	Total plasma concentration (units e.g., µM)	Any applicable bioanalytical method $K_{Pu} = \frac{C_T}{C_u}$	Assessment of total plasma concentration exposure

Chapter 4

Principles of Dose, Time and Flow Dependencies

4.1 What is this chapter about?

This chapter covers different aspects of nonlinear pharmacokinetics caused by changes in *dose*, *time* and *flow* and the resultant alterations in absorption and elimination. Specifically, this Chapter focuses on:

- establishing correct dose-concentration relationships in the design of pharmacodynamic or safety experiments, particularly when it is necessary to extrapolate the dose-concentration relationship outside the experimental range
- target mediated drug disposition *TMDD*, a special case in which nonlinear pharmacokinetics may be observed and encompass both non-linearities in distribution and elimination, as a consequence of target abundance, turnover and drug on/off binding rates.

In its simplest general form, PK/PD for a drug faithfully portrays the level of exposure in relation to pharmacodynamic effect. However, it may also be that exposure estimates are dissociated from pharmacodynamic responses because of concomitant drug effects on factors governing disposition, and/or that the actual pharmacodynamic mode of target interaction has repercussions on pharmacokinetic parameters. We have in our own practice encountered many concrete situations where the dose-concentration relationship is poorly established, or even lacking. To illustrate why it is important to take nonlinear kinetics into account, a practical example of the *dose-concentration* relationship is given. Figure 4.1 shows why a doubling of the dose may either result in a *more* than doubling of the plasma concentration (left) or *less* than doubling (right). Case A in Figure 4.1 may be due to saturation of the eliminating system, and Case B due to solubility, or other absorption limitations or to nonlinear plasma protein binding. Plasma protein binding has been covered in Chapter 3 and saturation of enzymatic capacities will be discussed in Section 4.2. Consequently a change in the dose may deviate from a direct proportionality between the plasma concentration and pharmacological response.

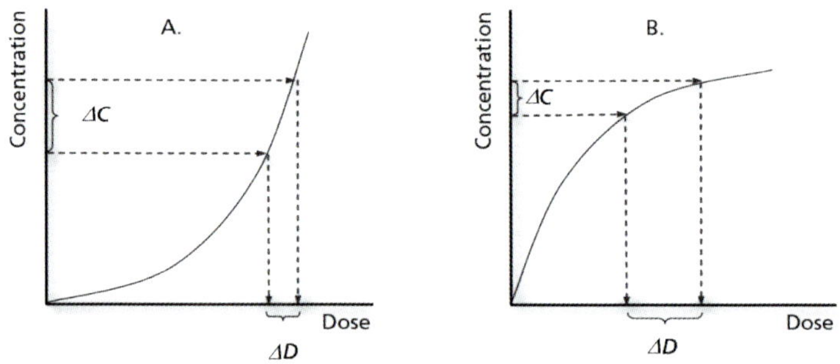

Figure 4.1
Schematic illustration of dose-concentration relationship of a capacity-limited elimination system (A) and a capacity-limited absorption system (B). Plot B may also represent the dose-concentration relationship when solubility of the test compound is limited.

A full understanding of the dose-concentration relationship allows the experimenter an insight as to why it can have a pronounced impact on the interpretation of the pharmacological responses. The dose-response and concentration-response relationships may have totally different shapes due to nonlinearities occuring between administered dose and the resulting plasma concentration. Table 4.1 lists a number of sources of commonly seen nonlinearities during absorption, distribution, elimination and turnover.

Table 4.1
List of nonlinear pharmacokinetic processes and their causes of nonlinearity

Pharmacokinetic process	System affected	Consequences
Absorption	Dissolution rate and solubility Transit time of the GI tract Transport across intestinal barriers	Less than proportional increase in *AUC*
Distribution	Saturation of plasma proteins	Total plasma concentrations behave nonlinearly
	Saturation of blood cell binding	Total plasma concentrations behave nonlinearly
	Saturation of tissue binding sites	Total tissue concentrations behave nonlinearly
Elimination	Saturable metabolism	Greater than proportional increase in *AUC*
Turnover	Metabolizing enzymes Target depletion	Can either be inhibited or induced resulting in increased and decreased parent and metabolite concentrations, respectively

Linear kinetics is defined as a proportional increase in the plasma concentration with an increase in dose. Pharmacokinetics is said to obey *first-order kinetics* when the rate of change in the plasma concentration is directly proportional to the concentration. Often linearity is assessed by means of the *dose-AUC* relationship. The most important feature of a linear system is that it obeys the law of superposition, which means that the dose-normalized concentration-time curves are superimposed lie on top of each other. In this sense the half-life is concentration-independent. In *nonlinear kinetics* the plasma concentration or *AUC* is no longer linearly proportional to the administered dose when clearance mechanisms display saturation. Consequently, a doubling of the dose may, for example, lead to tripling of the plasma concentration, or to little concentration change at all. The steady-state concentration may slowly increase or decrease over time due to inhibition or induction of the metabolizing enzymes. Transporters involved in either the absorption, distribution or elimination of the drug may also become saturated, and binding to various plasma proteins may be saturated. The rate of change of the plasma concentration is no longer directly proportional to the concentration in a nonlinear system. When the concentration increases, the rate of change levels off. Some nonlinear systems may behave as a linear system at low plasma concentrations. There are also nonlinear systems that become linear when plasma concentration or dose increases. Examples of such a behaviour are salicylic acid, and often antibodies.

4.2 Capacity-dependent kinetics

The expression *dose-dependent* kinetics is really more correctly referred to as *concentration-* or *capacity-dependent* kinetics. However, since *dose-dependent* kinetics in spite of its duality is such a well-established term, it will be used in the continued discussion unless *concentration-dependent* kinetics is needed for clarity.

The first examples of *dose-dependent* kinetics show that the extent of absorption may be due to poor solubility (Test compound X, Figure 4.2), saturable assimilation/uptake/transport processes (Vitamin B_{12} uptake, Figure 4.3) or capacity dependent metabolism (ethanol bioavailability, Figure 4.4). All three processes will have an impact on how to assess the *extent* of bioavailability.

For example, decreased solubility with increasing doses will show up as first-order absorption rates at low doses, and mixed first- and zero order when doses are increased (and solubility is exceeded). If the doses are further increased, the absorption rate limiting step will be zero-order and stays at zero-order as long as some of the drug substance has not dissolved. The data in Figure 4.2 indicate not only solubility-limited absorption for compound X but probably also a reduced gastro-intestinal motility elicited by this agent, resulting in an extended absorption over a 50 h time period. The transit time of the gastro-intestinal tract in dogs is about 20–24 h and the plasma peak concentration occurs at the end of this time for all doses except the lowest one (150 µmol/kg), possibly indicating lesser relative effect on gastric emptying by this dose. As another illustration, if the rate of ethanol absorption is decreased as the result of delayed gastric emptying (cf. also Figure 4.6 below), it will automatically lead to decreased ethanol absorption due to saturable extraction across the liver.

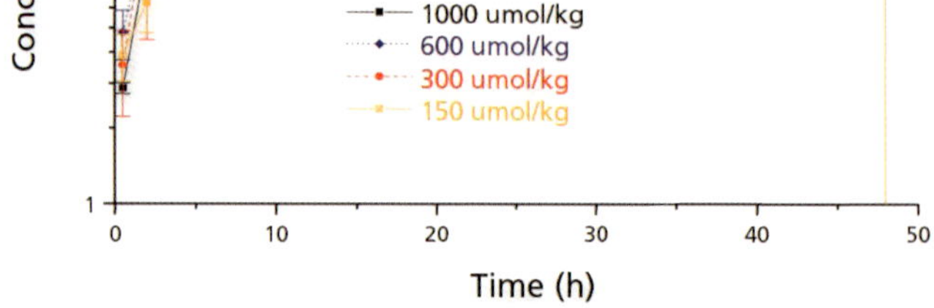

Figure 4.2
Semi-logarithmic plot of concentration-time data of compound X after oral gavage of 150, 300, 600 and 1000 µmol·kg^{-1} to Beagle dogs. The peak plasma concentration increases only 2-fold with a 7-fold increase in dose due to the poor solubility of compound X.

Yet an example where physiological processes markedly influence absorption is shown in Figure 4.3. Vitamin B_{12} is taken up almost to 100 % at a dose less than 1 µg. However, when the dose is increased to 500 µg, the fraction absorbed is close to only 1 %. For its absorption, this vitamin not only requires the presence of the so-called intrinsic factor (IF) protein, but also utilizes specific gastro-intestinal transporters. The data shown are consistent with a saturable absorption transport process limiting the uptake of Vitamin B_{12}.

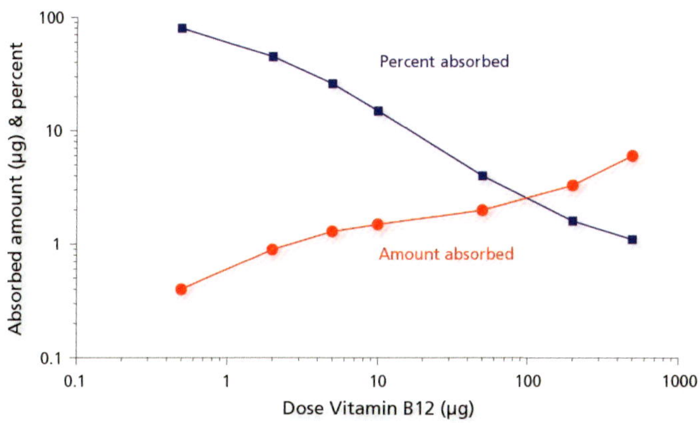

Figure 4.3
Double-logarithmic plot of dose of Vitamin B_{12} *versus* absorbed amount and percent absorbed dose. The uptake of Vitamin B_{12} occurs via a saturable gastrointestinal transport process.

A similar case is apparent for Vitamin C (ascorbic acid), megadoses of which are commonly self-prescribed to fight cold symptoms. However, the body does not fully utilize doses above 100 mg, and megadoses of Vitamin C have a marginal effect on both the actual amount absorbed and the peak plasma concentration. Consequently, even after a 100-fold increase in the Vitamin C dose, there is only an 8-fold increase in the amount absorbed (Table 4.2). A rise in Vitamin C dose from 100 mg to 2500 mg resulted in an modest rise in peak concentration from 12 to 15 mg·L^{-1}. Also from the pharmacodynamic point-of-view, there is little support for useful therapeutic efficiency of high Vitamin C doses for prophylaxis and treatment of common cold in the general population.

Dose (mg)	Per cent absorbed (%)	Amount in the body
100	100	100
1000	20	200
10000	8	800

Table 4.2
Dose of Vitamin C (mg), per cent absorbed (%), and amount of Vitamin C in the body (mg)

Another type of capacity-dependency is seen in the relationship between the concentration of ethanol and its bioavailability across the liver, which is simulated and depicted in Figure 4.4. With a plasma K_m value of about 30 mg·L^{-1}, the saturation of the ethanol kinetics begins to occur following a small dose. This relationship is typical for a system that obeys saturable first-pass metabolism. The bioavailability is low at low hepatic tissue concentrations (20 % at 1 mg·L^{-1}), but when the concentration rises so does the bioavailability (*e.g.*, 60 % at 100 mg·L^{-1}). Above 1 g·L^{-1}, the bioavailability is almost complete. By raising the absorption rate (concentrations in portal blood will rise and so will the exposure to the hepatocytes), the extent of bioavailability is automatically raised due to the saturable first-pass metabolism.

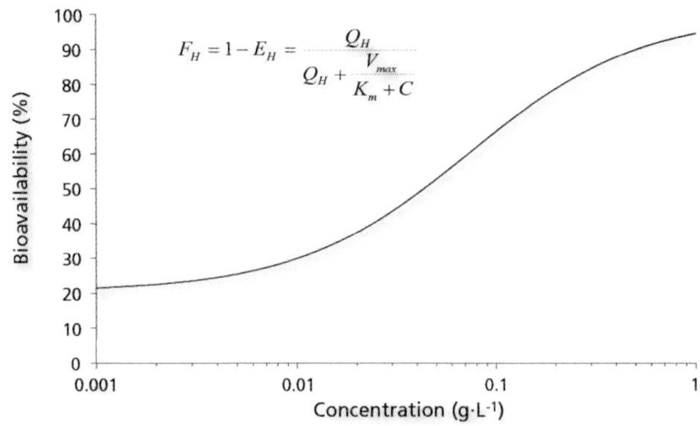

Figure 4.4
Semi-logarithmic plot of simulated concentration (g·L^{-1}) versus bioavailability (%) of ethanol in humans. The simulation is based on data obtained from Norberg *et al* (2000).

The consequences of a capacity-limited elimination *versus* a capacity-limited uptake process on the plasma concentration-time profiles of two different test compounds at three different doses are shown in Figure 4.5. The 100-fold increase in dose results in a greater than 100-fold increase in the peak plasma concentration when there is capacity limited elimination (Figure 4.5, left). The areas under the plasma concentration-time curves increase disproportionately to the increase in dose. However, the system strictly obeys capacity-limited absorption and the dose-normalized areas superimpose. As a result a nine-fold increase in dose only produces a 3-fold increase in the peak drug concentration, and linear pharmacokinetics prevail with a constant *dose-to-AUC* ratio with increasing doses.

Figure 4.5
Left: Semi-logarithmic plot of plasma concentration (nM) time data of test compound X after gavage to rats at 3, 30 and 300 µmol·kg^{-1}.
Right: Semi-logarithmic plot of plasma concentration (µg·L^{-1}) time data of test compound Y after gavage to rats at 10, 30 and 90 mg.

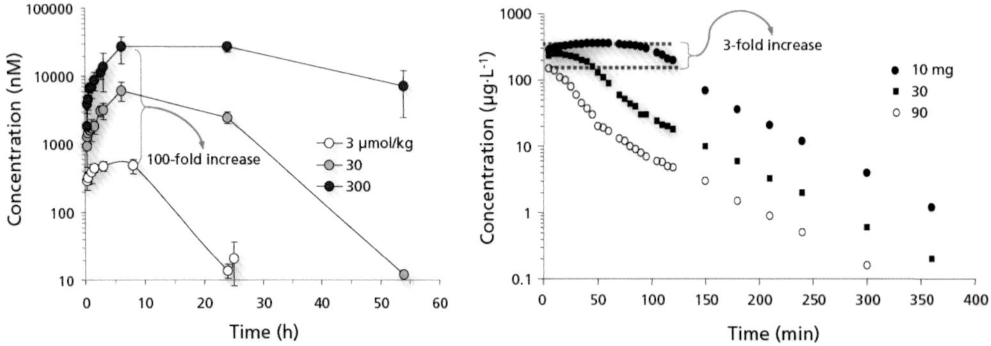

Figure 4.6 shows the plasma kinetics of ethanol in 10 human volunteers who received a standardized oral dose of 20 g ethanol after either fasting (left), or together with a meal (middle) or together with a meal and a dose of aspirin (right). The slower the absorption rate the lower is also the extent of bioavailability, as judged from the area under the plasma concentration-time curves. The slower absorption rate following the aspirin-induced reduction in gastric emptying, allows an extended slow rise in the plasma concentration-time course and thereby also a higher extraction of ethanol across the liver. This explains the common experience that pharmacodynamic effects of ethanol are much less salient when consumed in conjunction with food than under fasting conditions – a reflection of protracted absorption kinetics leading to a delayed and lower C_{max}.

Figure 4.6
Upper row: Semi-logarithmic plot of plasma concentration ($g \cdot L^{-1}$) time data of ethanol after a standardized dose of 0.3 g ethanol per kg to human volunteers at three different occasions. The left hand plot shows the ethanol kinetics in the fasting state, the middle plot together with a meal, and the right hand plot together with a meal and an a dose of aspirin.
Lower row: Semi-logarithmic plots of the same data. Note the increased variability when ethanol is taken together with a meal or a meal+aspirin. The combination with aspirin causes a pronounced delay in the gastric emptying and absorption of ethanol with a plasma concentration-time profile that peaks at about 100 min.

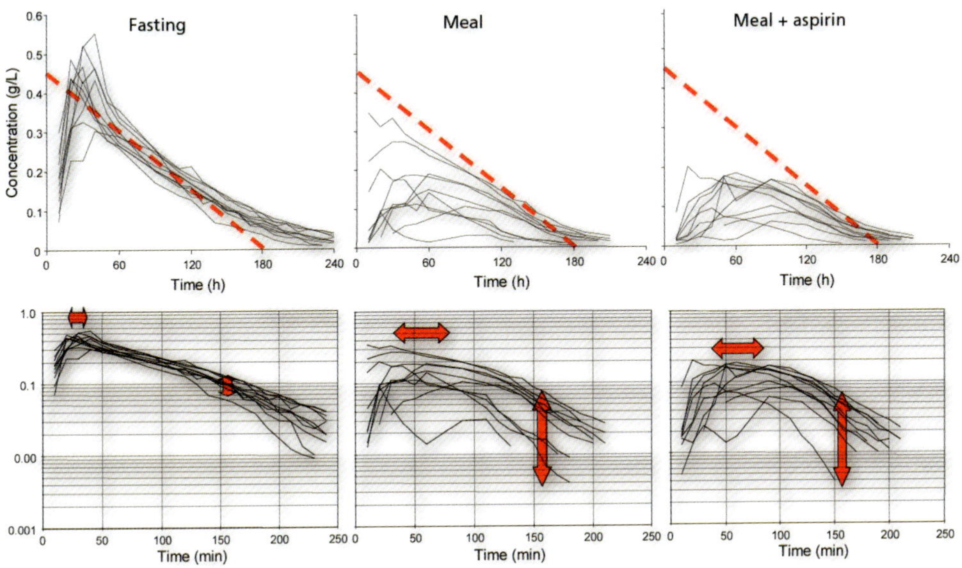

How can we ascertain when the pharmacokinetics are nonlinear? For a drug that obeys first-order kinetics, all dose-normalized concentration-time profiles superimpose (Figure 4.7, left). If the drug displays capacity-limited elimination, the principle of superposition does not apply (Figure 4.7, right).

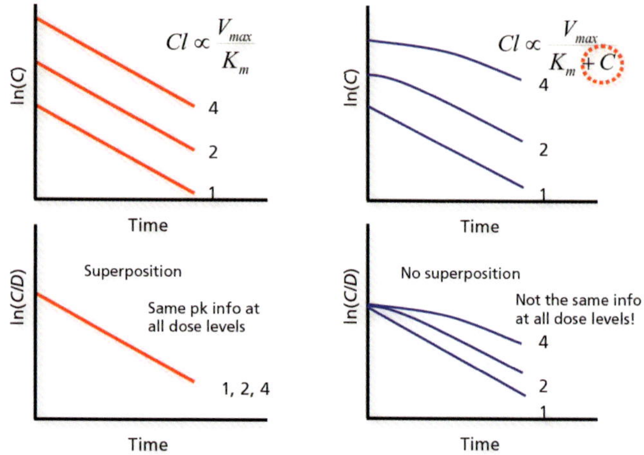

Figure 4.7
Left: Schematic illustration of three intravenous bolus injections in a system which obeys first-order disposition (upper left hand plot). Clearance is a concentration-independent parameter proportional to V_{max} divided by K_m. Dose-normalized plasma concentrations superimpose (lower left hand plot).
Right: Schematic illustration of three intravenous bolus injections of a system obeys nonlinear disposition (upper right hand plot). Clearance is a concentration-dependent parameter proportional to V_{max} divided by (K_m + C). Dose normalized plasma concentrations do not superimpose (lower right hand plot).

When the areas and concentrations cover a large enough range, the possibility of clearly comparing high and low concentrations becomes less practical. To circumvent this problem, the relative contribution of each concentration or area measurement is obtained by normalizing the area or concentration value by its corresponding dose. This is a commonly used approach, and is very useful if plotted *versus* dose (Figure 4.8). This type of exploratory data analysis is recommended in all kinetic analyses prior to any model derivation.

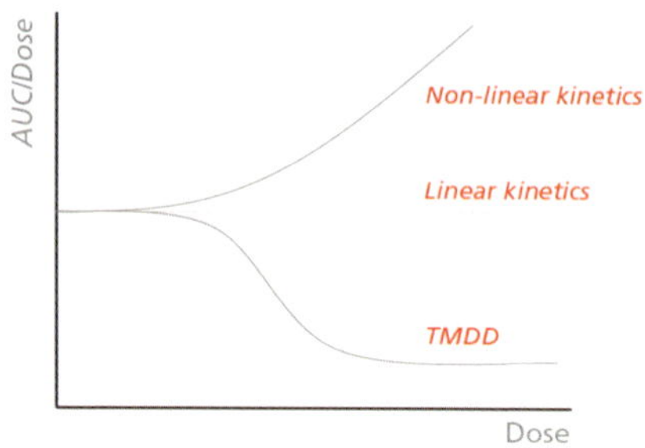

Figure 4.8
Schematic illustration of dose-normalized *AUC* versus *Dose*. The horizontal line represents a linear first-order system, the upper line a saturable system and the lower a typical Target-Mediated Drug Disposition (TMDD) system (cf. Section 4.5 below).

A consequence of capacity-limited elimination is that the principle of superimposition is no longer valid. This is shown in Figure 4.9. The capacity-limited system (left) is contrasted with a first-order system (right). Note that an increase in dose shifts the concentration proportionally for the first-order system. This is not the case for the capacity limited system where a doubling of the dose may lead to more than doubling of the plasma concentration. The more nonlinearity involved in the absorption and disposition, the more variability there will be in the data.

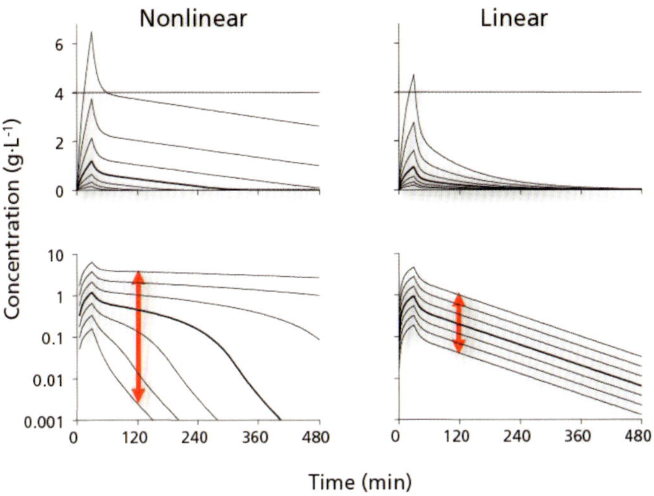

Figure 4.9
Linear (right hand panels) and nonlinear (left hand panels) kinetics. Note the large variability in concentrations as the dose increases for the capacity limited system (left) as compared to linear kinetics (right).

A capacity limited clearance is governed by the maximum metabolic rate V_{max}, the Michaelis-Menten affinity constant K_m and the drug concentration adjacent to the enzyme, C. The consequences of an increase in either the maximum metabolic capacity or the Michaelis-Menten constant are shown in Figure 4.10. An increase in metabolic rate V_{max} increases Cl and as a result decreases the area under the plasma concentration-time curve AUC. When the affinity between drug and metabolizing enzyme decreases, observed as an increase in K_m, the opposite effects on Cl (↓) and AUC (↑) are observed (Figure 4.10, right).

$$Cl = \frac{V_{max}}{K_m + C}$$

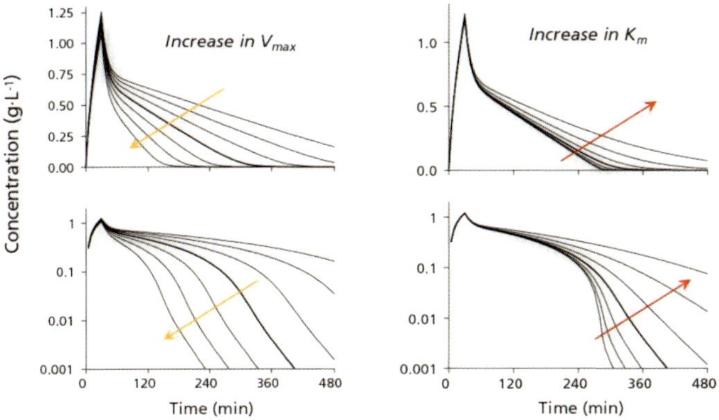

Figure 4.10
Left hand plots: Simulations of rapid intravenous injections followed by washout of a system that obeys saturable elimination and where the maximum metabolic rate V_{max} (capacity) has been increased.
Right hand plots: Simulations of rapid intravenous injections followed by washout of a system that obeys saturable elimination and where the affinity between drug and metabolizing enzyme decreases and thus Michaelis-Menten constant K_m has been increased. Upper row shows data on a linear concentration-time scale and bottom row plots on a semi-logarithmic scale.

The behaviour of the concentration-time courses shown in Figures 4.7 and 4.10 can be described mathematically by means of Equation 4.1

$$\frac{dC}{dt} = -\frac{Cl}{V} \cdot C = -\frac{V_{max}}{K_m + C} \cdot \frac{1}{V} \cdot C \tag{4:1}$$

Where V_{max}, K_m, V and C are the maximum eliminating capacity (amount·time^{-1}), Michaelis-Menten constant (amount·volume^{-1}), volume of distribution (volume) and plasma concentration (amount·volume^{-1}), respectively. The capacity-limited clearance (volume·time^{-1}) is expressed as

$$Cl = \frac{V_{max}}{K_m + C} \tag{4:2}$$

The larger the maximum eliminating capacity V_{max}, the higher is the clearance Cl, resulting in a lower AUC. The higher plasma concentration the lower is clearance, resulting in a larger AUC. The K_m parameter corresponds to the plasma concentration C at which clearance has decreased by a factor of 50 % of its maximum value V_{max}/K_m. The latter ratio is sometimes referred to as intrinsic clearance. The first-order term plasma half-life is commonly expressed as

$$t_{1/2} = ln\, 2 \cdot \frac{V}{Cl} \tag{4:3}$$

Under first-order conditions, clearance and $t_{1/2}$ are constant terms. If clearance obeys capacity-limited behaviour, then $t_{1/2}$ will be nonlinear. This may be appreciated by combining Equations 4:2 and 4:3 into Equation 4:4

$$t_{1/2} = \ln 2 \cdot \frac{V}{V_{max}} \cdot \left(K_m + C\right) \tag{4:4}$$

Then $t_{1/2}$ is only the tangent to the plasma concentration-time course at a specific concentration, and changes in proportion to the plasma concentration C. The practical meaning of $t_{1/2}$ is therefore only valid under first-order kinetics.

4.3 Time-dependent kinetics

The focus so far has been on *capacity-* or *concentration-*dependent kinetics caused by saturable absorption or elimination processes where the concentration of drug impacts on its own clearance. In such a case, clearance becomes concentration-dependent which is called *capacity-dependence*. However, when clearance varies over time due to a change (increase or decrease) in the amount of enzyme responsible for drug breakdown, it is called *time-dependent* kinetics. As Levy [1982] has pointed out *"The major distinguishing features between capacity (dose) and time dependency, is that the latter involves an actual physiological or biochemical change in the organ(s) of the body associated with the drug disposition parameters in question. For example, in time dependence of the auto- or hetero-induction type, the increase in drug intrinsic clearance results from an increase in amount of enzyme (e.g., in protein synthesis). However, in a typical Michaelis-Menten capacity (dose) dependency, drug clearance changes with concentration, and such a system should not be considered time-dependent simply because the values of pharmacokinetic parameters change with time. If that was a true time-dependent system, drug clearance should change with time while drug concentration is time invariant. It is still possible that capacity and time dependency exist simultaneously".*

A typical example of time-dependent kinetics is illustrated by the experiment where nortriptyline was given to normal subjects together with pentobarbital (Figure 4.11). Nortriptyline was administered orally as 10 mg every 8 h for a total period of 29 days. After 9 days, treatment with pentobarbital, was initiated and continued until 21.5 days from the start of nortriptyline dosing. Pentobarbital stimulates the synthesis or production of the enzyme responsible for clearance of nortriptyline. The concentration time course of nortriptyline before, during and after treatment with this enzyme inducer is shown in Figure 4.11, including the predicted time course of nortriptyline. *A* is time before induction. *B* is the induction period of the red bar and *C* is the period after the red bar. The more the enzyme is induced, the greater is the metabolic capacity and the higher is the metabolic clearance of nortriptyline, which leads to a lower nortriptyline plasma concentration. A schematic illustration of how pentobarbital induces the level of enzyme responsible of nortriptyline is shown in Figure 4.12.

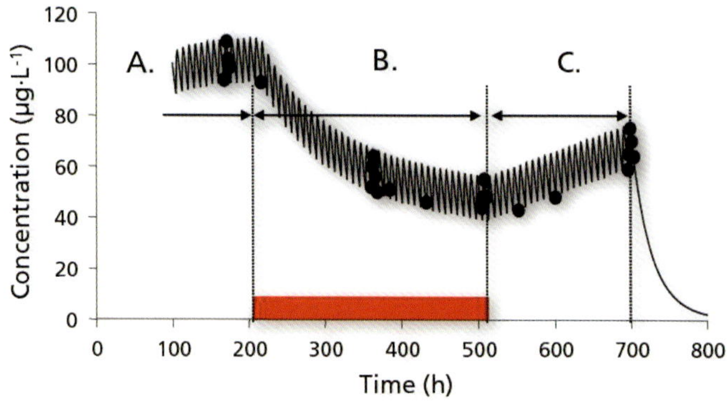

Figure 4.11
Concentration time course of nortriptyline before (A), during (B) and after (C) exposure to pentobarbital induction of the enzyme responsible for metabolism of nortriptyline. The red bar denotes the period of pentobarbital treatment. Note that the time to the induced steady-state is shortened due to an increased clearance and hence a shorter $t_{1/2}$. The time it takes for plasma concentrations to return back to the uninduced steady-state is now prolonged, as post-induction clearance of nortriptyline gradually decreases when enzyme levels return to normal, and hence $t_{1/2}$ becomes longer. Data are taken from von Bahr *et al*. (1998) and Case Study 20 in Gabrielsson and Weiner (2006).

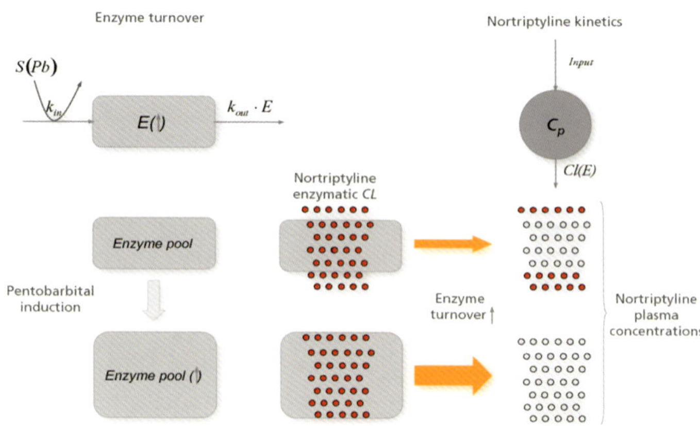

Figure 4.12
Schematic illustration of the turnover of the enzyme responsible for nortriptyline clearance. Stimulation of the production of enzyme by means of pentobarbital S(Pb) increases the enzyme level E, thereby increasing nortriptyline clearance *Cl(E)*, and hence lowering plasma concentrations of the compound (represented by filled dots in the Figure; open dots illustrate metabolized nortriptyline).

The pre-induction oral clearance was estimated to be 46 L·h^{-1} and the induced oral clearance was about 118 L·h^{-1}. This is equivalent to a corresponding increase in the enzyme level, which will have consequences for bioavailability and hepatic extraction. The average plasma concentration is halved within the observed time frame, which is consistent with a 100 % increase in the induced enzyme levels. The half-life, which governs the time to steady state, is determined by the volume of distribution and clearance. Since the half-life is shortened during induction, due to an increase in *Cl*, the time to establish the induced steady state is likewise shortened (Figure 4.13). When the inducer is removed, enzyme levels (and thereby *Cl)* gradually decrease and the nortriptyline half-life returns to its pre-induction value, which is longer compared to during the induction period. Therefore, the return of the induced state towards normal levels is significantly prolonged, also relative to the initial time required to attain steady-state.

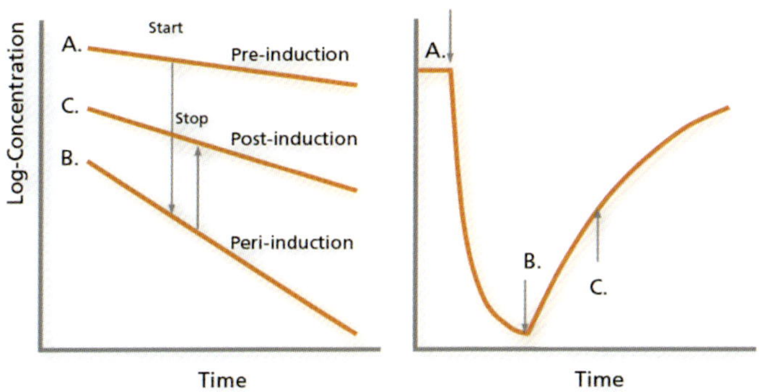

Figure 4.13
Schematic presentation of the log-concentration decline for a compound (such as nortriptyline) undergoing time-dependent changes in its elimination. A, represents the washout kinetics before induction; B, corresponds to the washout kinetics during induction; C, represents the washout kinetics some time after the exposure of the inducer has ceased. The time to a new steady-state is shortened during induction (half-life of drug is shortened) compared with the time after induction (half-life increases and returns to its pre-induction value). Note that the time scales in the left and right panels are not the same, and the concentration axis has arbitrary units. The time elapsed between A and B in the right hand panel is the period with the inducer.

Figure 4.14 illustrates three different scenarios of time-dependent clearance. The first is induction caused by the drug itself (auto-induction) or caused by another co-administered drug (hetero-induction, such as the described increase in nortriptyline clearance by means of pentobarbital). The second example illustrates time dependent inhibition, *e.g.*, due to the formation of an metabolite inhibitory to the degrading enzyme. The last case shows an oscillatory clearance caused by hormonal or food-induced changes in metabolic enzyme activity.

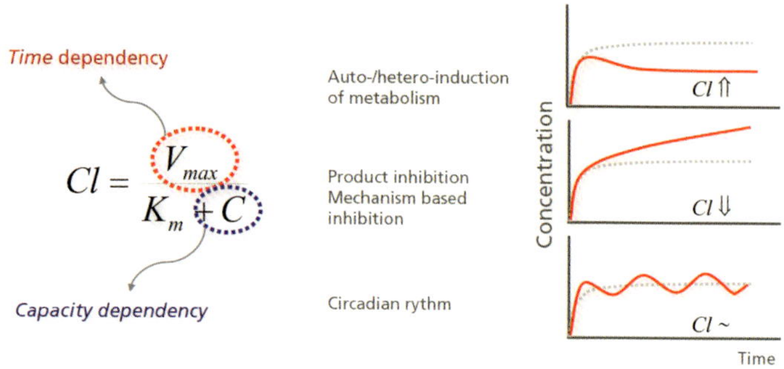

Figure 4.14
Schematic presentation of Capacity- and Time-dependent elimination processes.
Upper right: Clearance increases over time due to induction (grey curve normal state, and red curve induced state).
Middle right: Clearance decreases due to product inhibition (grey curve normal state, and red curve under the influence of product- or mechanism-based inhibition).
Lower right: Clearance oscillates over time due to activity changes driven by internal or external factors (grey curve in the absence, and red curve in the presence of time-dependent fluctuations in internal or external factors influencing enzyme activity).

There are situations, such as nonlinear tissue binding, *"where the volume of distribution and the apparent clearance is higher for the first dose as compared to subsequent doses, which could have implications for bioavailability and/or bioequivalence studies; where the apparent volume and clearance decrease with increasing doses; steady state infusion concentrations are reached more rapidly with increasing infusion rate; the zero time plasma concentration intercept of the true terminal phase is not dose proportional and is almost independent of dose"* (Cheung and Levy [1989]).

4.4 Combining Capacity, Flow and Time-dependent kinetics

The previous sections have covered *capacity* (Section 4.2) and *time* (Section 4.3) dependencies separately. This information is now extended to include *flow*-dependent kinetics and thereby combine the three, using ethanol as the example. Variability in the absorption and disposition of ethanol has important consequences in clinical and forensic sciences. At blood ethanol concentrations below 10 $mg \cdot L^{-1}$ the kinetics is first-order and consequently ethanol displays perfusion rate-limited elimination. Clearance is flow dependent due to a high hepatic extraction ratio which is then greater than 0.7. At relatively low blood ethanol concentrations of less than 50 $mg \cdot L^{-1}$, the metabolism of ethanol starts to display capacity limited elimination. The combined effects of *capacity* and *flow* on blood ethanol clearance are demonstrated in Figure 4.15.

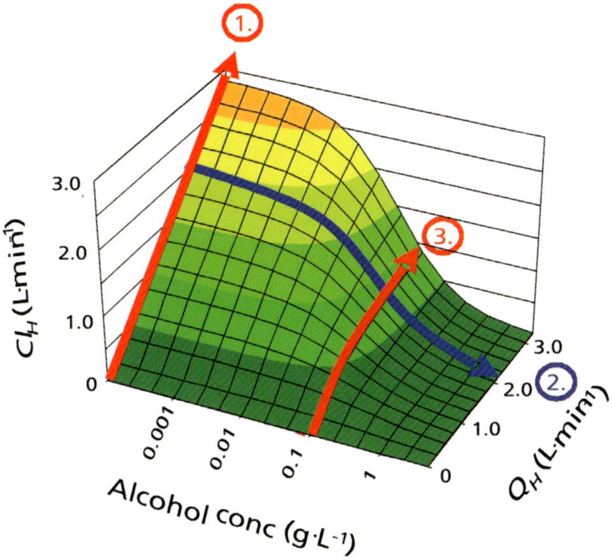

Figure 4.15
3-D plot of ethanol blood clearance (L·min^{-1}) versus alcohol blood concentration (g·L^{-1}) and hepatic blood flow (L·min^{-1}). Position ① shows a perfusion-limited clearance, ② saturable clearance, and ③ impact of a change in hepatic blood flow for a pertially saturated clearance.

The blue line in Figure 4.15 shows how ethanol clearance decreases at a fixed hepatic blood flow when the ethanol concentration increases – ethanol clearance goes from perfusion rate limitation (high extraction ratio) at low concentrations (①) to capacity rate limitation (nonlinear and low extraction) at high concentrations (②). A consequence of capacity limited elimination is that the principle of superposition is no longer valid. Ethanol is also being eliminated via the urine. At low ethanol blood concentrations only 1–2 % of the dose is renally excreted as ethanol. At higher ethanol concentrations, metabolic clearance becomes more saturated and a larger fraction of ethanol is excreted unmetabolized via the urine (about 10 %). In this higher concentration range, ethanol also decreases the release of vasopressin and thereby through increased urine volumes adds to increased urine volumes, hence adding to the greater renal elimination of unmetabolized ethanol. The combination of saturable metabolic clearance and low and variable renal excretion of ethanol makes it more difficult to backtrack exposure to ethanol in the blood when using urinary data which therefore should be avoided.

The inducible isozyme *CYP2E1* is also involved in ethanol elimination. However, it is unclear whether this enzyme actually contributes significantly to the overall elimination of ethanol. Nonetheless, from a conceptual point of view it is interesting to note that low ethanol concentrations do seem to stabilize *CYP2E1,* and therefore slow down the loss of enzyme. At higher ethanol concentrations additional induction is thought to be due to increased synthesis of *CYP2E1*. No study has yet specifically addressed the quantitative aspects of the multiplicative effect of a decreased loss combined with an increased synthesis rate of *CYP2E1*. This variability in ethanol kinetics stems from a combination of both genetic and environmental factors, and from the nonlinear nature of ethanol disposition itself due to *capacity, time*, and *flow* dependent phenomena. This seriously hampers making any prediction of the blood alcohol concentration on the basis of a given drinking pattern. Consider Figure 4.16; some of the subjects show a very rapid absorption with an ethanol peak concentration at the first sampling time while other subjects show a pronounced delay in absorption, with an ethanol peak concentration at 2 h. This large variability in blood ethanol concentrations needs to be carefully considered in forensic casework. It is also notable that there is a 30-fold variability in the terminal ethanol concentration at seven hours, where some of the individuals have reached the first-order linear concentration range.

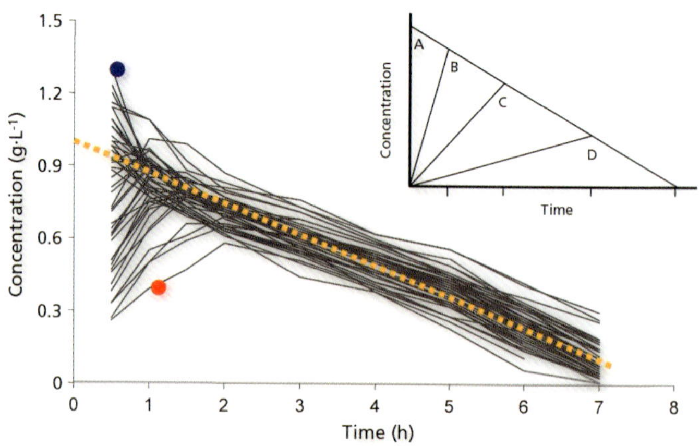

Figure 4.16
Concentration-time data of ethanol after a standardized dose given to 48 male volunteers given 0.6 g ethanol per kg body weight under standardized conditions. The inset shows a semi-logarithmic presentation. The dashed horizontal red line symbolizes the average K_m value of Norberg *et al* [2003]. The blue and red spots show concentrations of ethanol in two subjects. The inset plot shows a simulation of the consequences of zero-order elimination. Case A illustrates a bolus injection, B a short zero-order infusion, and C and D more extended infusion intervals. All doses are the same.

The *rate* and *extent* of absorption of ethanol from the stomach and small intestine into the systemic circulation will vary as a consequence of capacity, time, and flow limited elimination and both are highly dependent on the actual hepatocyte exposure to ethanol. Studies on the extent (bioavailability) of ethanol absorption based on *AUC* may therefore easily become misleading due to the operation of capacity-dependent (nonlinear) elimination. In cases of capacity-dependent first-pass, the rate of uptake of ethanol from the stomach and small intestine will confound any interpretation as to the reason for the extent of absorption. For similar oral doses of ethanol, the subject with the slowest *rate* of uptake will also show the lowest *extent* of absorption. This is illustrated by the individual identified by the red spot in Figure 4.16. The inset plot shows a simulation of the consequences of zero-order elimination. Case A illustrates a bolus injection, B a short zero-order infusion, and C and D more extended infusion intervals. The amount of ethanol in the body at the end of each infusion is the difference between the administered dose and the amount eliminated during the infusion period. Consequently, the concentration at the end of the infusion is the same as that expected at that time following the bolus injection. Note that the slower the input rate, the smaller is *AUC*, and the lower is the peak concentration.

4.5 Target mediated drug disposition (TMDD)

Target mediated drug disposition TMDD captures the capacity-limited binding and saturation of a drug-receptor complex, where a significant portion of the drug binds to its target such that this interaction influences the disposition of the drug. Figure 4.17 (left) depicts a schematic description of a generic turnover model for ligand binding to receptors circulating, the blood or plasma. The ligand may represent any type of molecule, such as antibody, protein, peptide or even a small molecule, the disposition of which is mediated by its target.

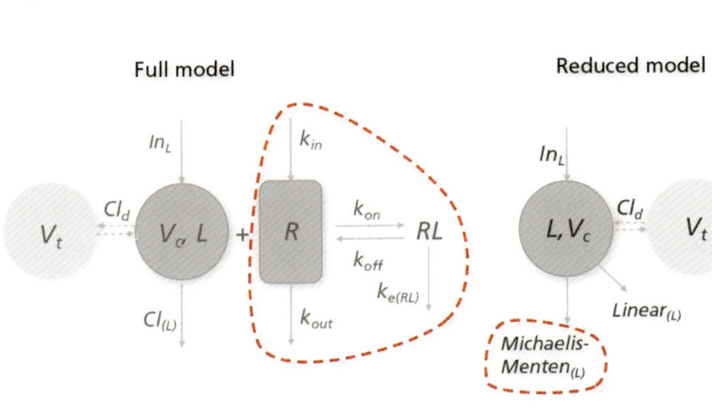

Figure 4.17
Left: Schematic illustration of the full Target-Mediated Drug Disposition model. The drug (ligand) distributes non-specifically over a central and peripheral compartment, being governed by its central V_c and peripheral V_t volume and inter-compartmental distribution term Cl_d. The drug (ligand) also reacts with the target (receptor) via a second order process and forms the receptor-ligand complex RL. The ligand can be released from the complex and return to the ligand pool (V_c) or the complex can be irreversibly removed from the system; described by $k_{e(RL)}$. We assume that the target is produced via a zero order production k_{in} and lost via a first-order process k_{out}. The ligand can also distribute into a non-specific tissue binding space (V_t). **Right**: The reduced TMDD model is used in the analysis of the Bauer *et al* (1999) data in Figure 4.18. The circled part of the full model is simplified to a Michaelis-Menten term in the reduced model.

Digitized data of Bauer *et al* (1999) can be used as an introductory example of *TMDD*. In Figure 4.18 data are shown on observed and model predicted concentration-time courses of anti-CD11a antibody hu1124 in psoriasis patients is shown, following injections of 0.1, 0.3, 1.0, 3.0 and 10 mg·kg^{-1}.

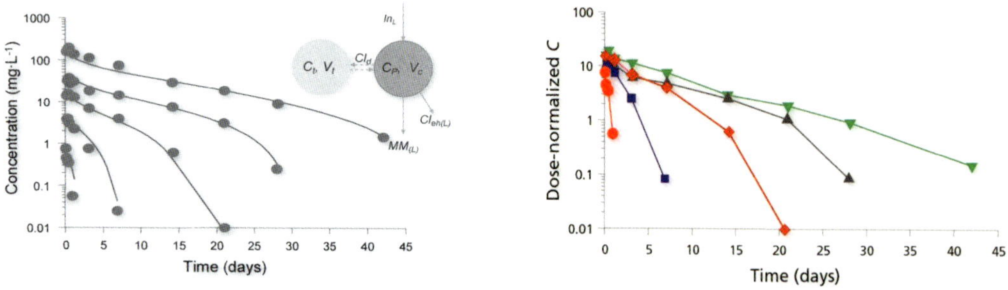

Figure 4.18
Left: Semilogarithmic plot of Antibody hu1124 concentrations versus time in human psoriasis patients after rapid intravenous injections of 0.1, 0.3, 1.0, 3 and 10 mg·kg^{-1}. Data digitized from Bauer *et al* (1999). The inset shows the two-compartment model with parallel Michaelis-Menten and first-order clearance of the antibody.
Right: Dose-normalized plasma concentrations versus time displaying the typical lack of superposition with increasing doses.

Due to the lack of data on receptor and receptor-ligand complex, the full model has to be re-parameterized to a reduced model (Figure 4.17, right). This allows fitting a two-compartment model with parallel first-order and Michaelis-Menten clearance terms. The model used to fit the experimental data in Figure 4.18 is shown as the *reduced* model in Figure 4.17.

Levy wrote in a seminal paper on pharmacological target-mediated drug disposition that *"Drug bound to target sites can persist much longer than indicated by its apparent plasma* $t_{1/2}$ *(i.e., the* $t_{1/2}$ *of elimination at concentrations higher than the true terminal phase, which parallels that of drug at target sites) but it is reversible and can be mobilized readily by competitive displacers. If biotransformation and excretion are apparent first-order processes and if drug binding to plasma proteins is either negligible or independent of concentration, then: 1) For single doses not preceded by other doses (even days or weeks earlier, depending on the drug), the apparent volume of distribution and total clearance decrease with increasing dose but reach a limiting value. This is most apparent with doses small enough that a significant fraction is bound to pharmacologic target sites. Bioequivalence and bioavailability studies may therefore exhibit order effects. 2) After the first dose, the single- and multiple-dose kinetics are first order (dose proportional). 3) Constant-rate infusion (particularly at relatively low rates) in subjects without previous exposure to the drug results in plasma concentration profiles having an initial lag-phase because of acquisition of the drug by the target sites in tissues. This has been shown in humans with the ACE inhibitor perindoprilat. (Conversely,* *if all or most of the high affinity binding sites are in plasma, the initial drug accumulation rate in plasma will be higher than indicated by the drug's apparent biologic $t_{1/2}$).*"

Considerable reserve capacity was found for coumarin-sensitive reductase, which may be one of several explanations why at least 70 % of the hepatic vitamin K_1 2,3-expoxide reductase activity must be inhibited before the vitamin K-dependent carboxylation of the clotting factors becomes compromised and prothrombin time changes occur. Another source of concern related to *TMDD* is the pharmacological consquences of the mobilization of drug bound to high-affinity sites in tissues, such as that described for the aldose reductase inhibitor imirestat, used to treat diabetic cataract. Levy's question that "*Can such mobilization produce an unexpected and undesired resurgence of a pharmacologic effect?*" merits investigation. Another example is the long-lasting abolition of additional hypothermic effects upon return to the baseline seen after a single subcutaneous dose of 300 μmol·kg^{-1} of clomethiazole to rats. Almost a month of recovery was needed after the appearance of the normal hypothermic effect of clomethiazole was seen again. For review of TMDD see Levy (1994) and Peletier and Gabrielsson (2009).

4.6 How is nonlinear drug disposition integrated with PD?

This chapter has discussed how dose (concentration) and time-dependent processes affect the exposure to a drug. This is of importance when pharmacological responses are measured over time and analyzed. We highlight the dissemination of nonlinear kinetics from nonlinear dynamics. An erroneously predicted exposure profile should not confound the pharmacodynamic model or parameters (*e.g.*, E_{max}, EC_{50}). How do we know when the pharmacokinetics of a compound is nonlinear? The most robust way to assess nonlinear pharmacokinetics is to dose-normalize the concentration-time course obtained from two or more doses. If the dose-normalized concentration-time profiles do not superimpose, we probably have some contribution from nonlinear absorption or disposition. To fully study the impact of the nonlinearity, concentration-time data of two or more dose levels are needed, and not only following extravascular dosing. An intravenous dose is needed to rule out the potential contribution from nonlinear disposition. Ideally, data from repeated dosing are also needed, coupled to *ex vivo* measurements of the plasma protein binding. Remember that clearance is not a single constant parameter in capacity-dependent disposition. Clearance is then a combination of maximum metabolic rate V_{max}, the Michaelis-Menten parameter K_m and the variable plasma concentration C. Recall from Chapter 3 that a nonlinear free fraction may cause a nonlinear behaviour in the total plasma concentration C *in spite* of the fact that the unbound concentration obeys linear kinetics. That is the primary reason why we also want to complement total plasma concentrations with *ex vivo* information on plasma protein binding. Changes in f_u affects C but not C_u. Table 4.3 summarizes some pivotal parameters related to nonlinear pharmacokinetics, how they are measured and how to apply them.

Table 4.3
Summary table of some pivotal disposition parameters V_{max}, K_m and f_u.

Parameter	What is it?	How is it measured?	How do you apply it?
V_{max}	Maximum metabolic capacity	Non-linear regression of concentration-time data following two or more dose levels	$Cl = \frac{V_{max}}{K_m + C}$
K_m	Michaelis-Menten constant and measures primarily the affinity between susbtrate and enzyme	Non-linear regression of concentration-time data following two or more dose levels	$Cl = \frac{V_{max}}{K_m + C}$
f_u	Free fraction	$f_u = \frac{C_u}{C} = \frac{1}{1 + K_a \cdot n \cdot [P]}$ f_u is often measured experimentally in equilibrium dialysis as the unbound-to-total concentration	f_u is often used to calculate unbound plasma concentrations from total, assessment of PK (Cl_u) and PD (EC_{u50}) parameters and to estimate safety margins

Figure 4.19 shows a range of scenarios that may be of interest for practicing pattern recognition. These scenarios are all taken from real life Case Studies, but they are intentionally simplified for the purpose of clarity.

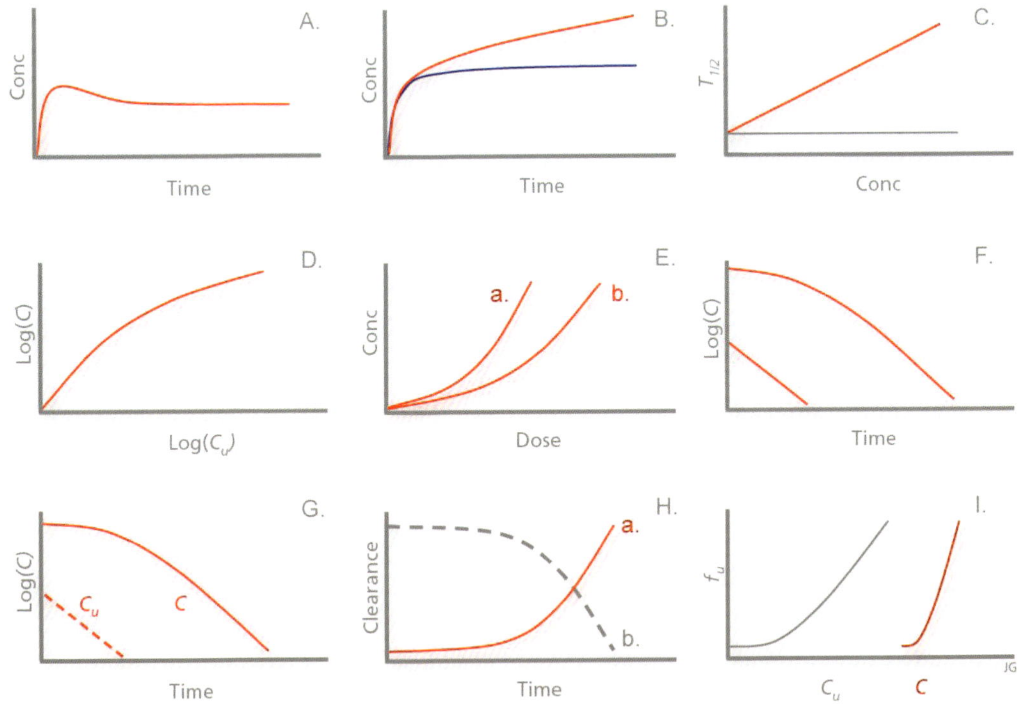

Figure 4.19
Schematic illustration of different nonlinear pharmacokinetic concentration patterns.

Plot A shows the concentration time-course after a constant-rate intravenous infusion with a time-dependent clearance. Clearance increases over time and reaches a new higher value (induced state), which is the reason why concentration initially increases and then decreases slightly. This situation is similar to the pentobarbital induced alteration in the elimination of nortriptyline.

Plot B shows two concentration-time courses after constant-rate intravenous infusion with a constant (time-independent) clearance (blue curve) and a time-dependent decrease in clearance (red curve).

Plot C shows how the half-life is independent of concentration (grey horizontal curve) and dependent of concentration (red curve). The mathematical relationship of the concentration-dependent half-life is given by Equation 4:4.

Plot D shows the relationship between unbound concentration and total plasma concentration for a system with saturable binding to plasma proteins. Since total concentration is actually a ratio of the unbound concentration-to-free fraction, and since free fraction increases with an increase in unbound concentration, the total concentration will not rise as steeply within a limited unbound concentration interval.

Plot E shows two dose-concentration courses exhibiting nonlinear kinetic properties. Curve 'a' denotes a relationship with a lower maximum metabolic rate V_{max} than Curve 'b'.

Plot F shows a nonlinear concentration-time relationship after two different intravenous bolus doses. The lower curve falls in the first-order linear concentration range whereas the upper curve displays typical capacity limited behavior.

Plot G shows the unbound and total concentration-time courses after a bolus dose. Since the total concentration may be viewed as a ratio of the unbound concentration-to-free fraction and saturable plasma protein binding is assumed to occur, total concentration will display a nonlinear disposition pattern, which is not due to capacity limited elimination but rather nonlinearities in PPB. Compare the upper curves of graphs F and G. Unless free fraction is known, it is difficult to discriminate between capacity-limited elimination (Plot F) and nonlinear plasma protein binding (Plot G) when only total plasma concentrations are measured.

Plot H shows time dependencies of clearance. Curve 'a' illustrates a system undergoing induction and Curve 'b' a system showing product inhibition.

Plot I illustrates how free fraction changes with unbound concentration when the number of binding sites on the protein starts to get saturated. The red curve shows how the free fraction is correlated to total plasma concentration (actually the ratio of unbound concentration-to-free fraction). Note the difference in steepness between the two curves.

Chapter 5

Principles of Rapid Concentration-Response Equilibria

5.1 What is this chapter about?

This chapter deals with concentration-response-time relationships where the equilibrium between plasma concentration and response occurs rapidly and no temporal differences can be observed between the two. Assuming a rapid equilibrium between the plasma concentration and the pharmacological effect then a behaviour illustrated by the left hand graphs of Figure 5.1 occurs. For each plasma concentration there is only one response. The same plasma concentration at two different time points results in the same response independently of whether they are rising or falling. The response is not lagging after the plasma concentration. There are no temporal differences between the two. That is, equilibrium occurs immediately. In contrast, the right hand graphs show temporal differences between concentration and response when plotted *versus* time (upper) and concentration *versus* response (lower) when plotted in time order (counter-clockwise hysteresis). This phenomenon is further discussed in Chapter 6.

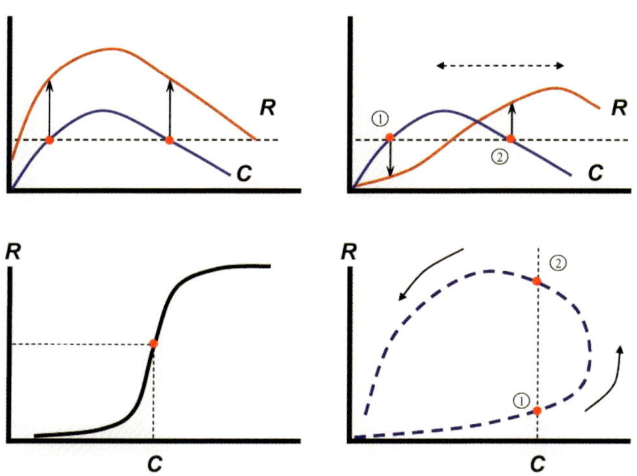

Figure 5.1
Left column: Schematic illustration of a rapid plasma concentration-response relationship. The concentration-response relationship follows the same trajectory up with increasing concentrations and down with decreasing concentrations.
Right column: Schematic illustration of a temporal difference between the plasma concentration-time and response-time courses. When concentration is plotted against its corresponding response in time order a hysteresis or a loop is observed.

The concentration-response relationships encountered are also applicable to the steady-state situation for most pharmacological effects, where time is not an issue. The differences between *dose-response* and *concentration-response* will be emphasized, as will be how one or the other changes an appreciation of the potential of a drug. A demonstration of how to apply the utility function within the context of the concentration-response will also be made. The plasma drug half-life is discussed from a pharmacodynamic perspective and the idea that the response halflife concept will be examined. The pharmacodynamic model gallery is examplified by five Case Studies using the *linear-* (Section 5.4), *exponential-* (Section 5.5), *log-linear-* and *Hill-* (Section 5.6), and *binding* model (Sections 5.7 and 5.8).

Potency, efficacy, steepness and selectivity based on the concentration-response relationship are discussed. *Potency* is defined as as the plasma concentration at which the drug-induced effect is 50 % of its drug induced maximum value. Potency has units of concentration (*e.g.*, $\mu mol \cdot L^{-1}$). *Efficacy* is the drug-induced response and its units depend on the measured response (arbitrary units). *Selectivity* is a greater proportion of the primary, desired, effect compared to secondary and/or negative effects and varies with concentration.

5.2 Why concentration-response instead of dose-response?

The comment that the *dose-response* relationship of a compound demonstrates its 'potency' has often been heard. To start with, *'dose'* should be avoided as a measure of *'potency'*. Potency is a measure of the affinity between a compound and a target, such as a receptor. Potency is therefore related to the thermodynamic activity, which is dependent on the local target concentration of the test compound. A dose, on the other hand, is equivalent to the amount administered and not to the active target concentration. The same dose of two compounds may give rise to totally different target concentrations depending on bioavailability, transporters and clearance although their affinities for the receptor are the same. Therefore, '*dose*' as a measure of potency should generally be abandoned.

The plasma concentration-response relationship may exhibit a different shape than the dose-response relationship, because concentration may not change in proportion to a change in dose (Figure 5.2, bottom graph). Consequently a doubling of the dose may lead to a minute change in the plasma concentration. However the opposite case may also be true because of the nonlinear relationship between dose and concentration as shown in Figure 5.3.

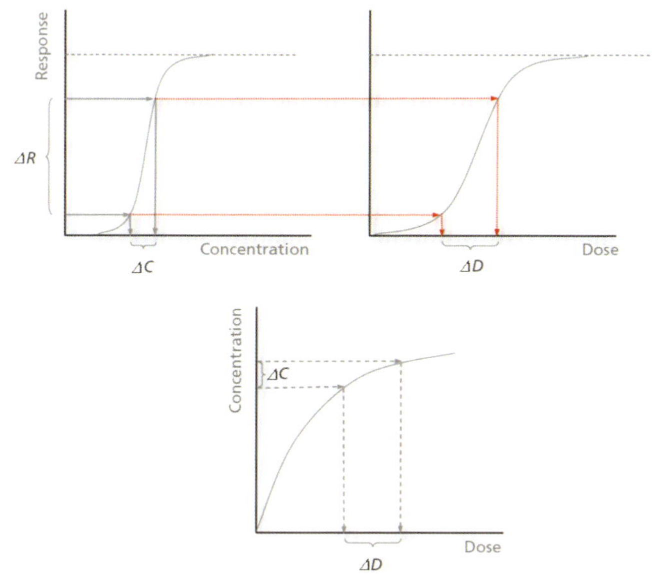

Figure 5.2
The concentration-response (upper left), dose-response (upper right) and dose-concentration relationship. Note the difference in steepness of the curves in the upper row. The shape of the dose-concentration curve may be due to solubility limitations at dosing site, capacity-limited absorption or nonlinear plasma protein binding.

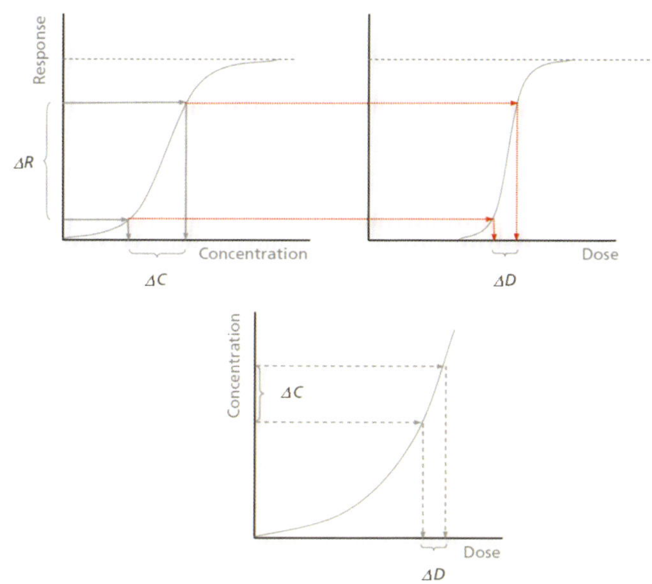

Figure 5.3
The concentration-response (upper left), dose-response (upper right) and dose-concentration relationship. Note the difference in steepness of the curves in the upper row. The shape of the dose-concentration curve may be due to capacity limited elimination or nonlinear plasma protein binding.

Moreover, hugely misleading conclusions may be drawn if comparing potencies of compounds with significantly different molecular weights based on *dose*-response curves. As already outlined in the preceding discussion, the plasma concentration-response relationship is therefore the measurement of choice because it gives a better idea about the effective concentration range, steepness, efficacy and selectivity between wanted and unwanted effects. The primary concentration-response (left grey curve) and unwanted concentration-response (right blue curve) are plotted in the same graph in Figure 5.4. There is higher selectivity at the low concentrations and lower or lost selectivity at higher concentrations. Note that selectivity is not a constant term, but varies with concentration. Another way of demonstrating the selectivity between primary and secondary (safety, toxicity) effects is to look at the utility curve. In that case there is higher selectivity between the primary and secondary effects on the upswing of the utility curve. When the utility curve starts to decline, the selectivity becomes less and less. Efficacy is maximized by aiming at the ascending portion close to maximum of the utility curve.

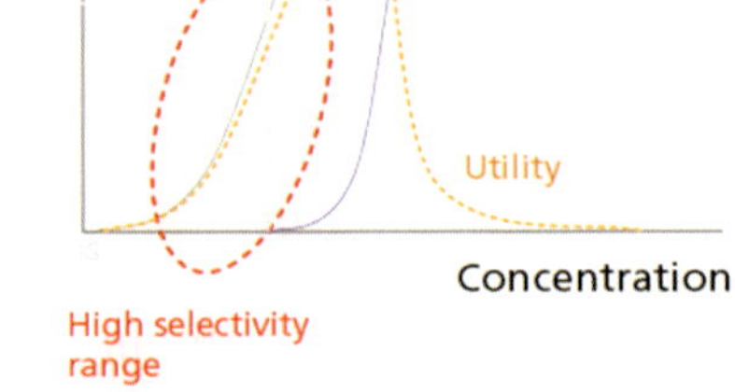

Figure 5.4
The concentration-primary response (grey curve, left), concentration-secondary response (safety marker, blue curve, right) and utility relationship (dashed grey curve). Note the difference in selectivity, which is a matter of concentration. The vertical arrow indicates the concentration at 50 % of maximal drug-induced effect. The dashed red circle shows the high selectivity portion of the primary concentration-response relationship.

The impact of different bioavailabilities on the assessment of pharmacological data is given by the real life example of a dose-response graph (left) and concentration-response graph (middle and right) of compounds A and B after oral administration at different doses (Figure 5.5). The oral availability of compound A is more than 95 % whereas for compound B it is less than 5 %. The left hand graph gives the impression that compound A is more potent than compound B. However, this apparent difference can simply be explained by the different bioavailabilities of A and B. Using a concentration scale both compounds exhibit similar potencies of about 10–15 nM. In fact, looking then at the relevant unbound concentration (right-hand graph), compound Y even tends to be somewhat more potent than compound X in this experimental setting.

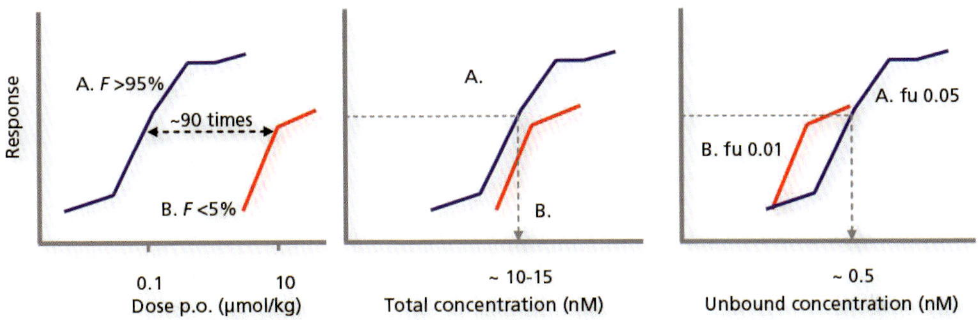

Figure 5.5
Semi-logarithmic dose-response (left plot) and concentration-response (middle plot) plots of compounds A and B in an *in vivo* rabbit model. The bioavailability of compounds A and B are >95 % and <5 %. The right hand plot shows the same response data plotted versus an unbound concentration scale. Note how the two compounds switch places and compound B becomes the more potent one. F and f_u denote the oral bioavailability and free fraction in plasma, respectively.

In answer to the initial question "Why *concentration-response* rather than *dose-response*?", it should be evident from the above discussion that dose-response relationships say nothing about what is thermodynamically active, nor do they necessarily give a correct shape of the concentration-response relationship due to pharmacokinetic nonlinearities and will, consequently, jeopardize the assessment of safety margin. There is still a misuse of the expression '*dose-response*'. It is common to hear '*dose-response*' in spite of the fact that most scientists actually mean '*concentration-response*'. Why is the former used when the latter is actually meant? The same inconsistency occurs between free or unbound plasma concentrations and the free fraction in plasma. The unbound concentration C_u has concentration units, for example, $\mu g \cdot L^{-1}$ or μM. The unbound concentration is determined by the oral dosing rate and unbound clearance. The free fraction is the ratio of unbound-to-total concentration and has no unit of measurement. The free fraction is determined by the affinity between drug and plasma protein, the number of binding sites and the concentration of the binding protein. Whenever different compounds, species, rates and routes are compared, the concentration-response relationship is required and it is even better to use the unbound concentration-response relationship as demonstrated in Chapter 3.

5.3 Multiple dose pharmacodynamics

Chronic medical conditions require chronic dosing, and generally drugs are prescribed to be taken at a fixed dose and time interval. The consequence of this kind of administration, is that the plasma concentration and amount of drug in the body will reach steady-state with a peak-trough profile, the magnitude of which depends on the actual dosing interval and PK properties. Table 5.1 shows the therapeutic plasma concentrations for a selected number of compounds.

Table 5.1
Selected compounds and their therapeutic plasma concentrations, clearance, volume of distribution, free fraction, half-life and daily dose

Compound	Disease Treatment	Therapeutic concentrations (μM)	CL (mL·min^{-1}·kg^{-1})	V_{ss} (L·kg^{-1})	f_u (%)	$t_{1/2}$ (h)	Dose (mg)
Cyclosporin	Organ transplantation	0.13–0.34	5.3	1.3	7	5.6	
Digoxin	Cardiac dysfunction	0.8–3 nM	130	440	3	39	
Felodipine	Antihypertensive	8±5 nM	12	10	<1	5	2–10
Gentamicin	Gram-negative infection	7–21	90	0.31	>90	2–3	
Lithium	Bipolar disorders	0.4–1.4	3.35±0.11	0.66±0.16	-	2.2	
Nitroglycerine	Antianginal	1.2–11 ng·ml^{-1}	230±90	5.3±1.2	-	2–4 min	
Nortriptyline	Depressive disorder	0.2–0.6	7.2±1.3	3.2±1.4	8	31±13	
Omeprazole	Ulcer	-	7.5±2.7	0.34±0.09	5	0.7±0.5	20–40
Phenytoin	Epilepsy	30–60	Conc -dependent	0.64±0.04	10	Conc -dependent	
Quetiapine	Schizophrenia, bipolar disorder, depressive disorder	-	-	10±4	17	6–7	150–300
Theophylline	Asthma and chronic obstructive disorder	30–100	0.65±0.2	0.5±0.16	40	90±2.1	
Valproic acid	Epilepsy, bipolar disorders	280–690	0.11±0.02	0.22±0.07	7	14±3	
Warfarin	Thromboembolic diseases	3–13	0.045±0.024	0.14±0.06	1	37	

It should be recalled that following an intraveous bolus dose the decline often displays a mono-exponential decline in a semi-logarithmic plot of concentration-time (Figure 5.6). The pharmacological response, on the other hand, stays relatively flat when concentration exceeds EC_{80}, but declines approximately linearly with time within the EC_{80} to EC_{20} range because it is proportional to the log-concentration, and then declines exponentially below EC_{20}. With respect to the pharmacological response there is nothing like a half-life of the effect unless one is far below EC_{20}. If the concentration stays in the EC_{20}-EC_{80} concentration range, the response varies in direct proportion to the logarithm of the plasma concentration. Although the plasma half-life of *e.g.* felodipine is about 5 h (*cf.* Table 5.1), a daily dose of 2 to 10 mg suffices, because the therapeutic concentrations reside in the vicinity of EC_{50} for a large portion of the dosing interval.

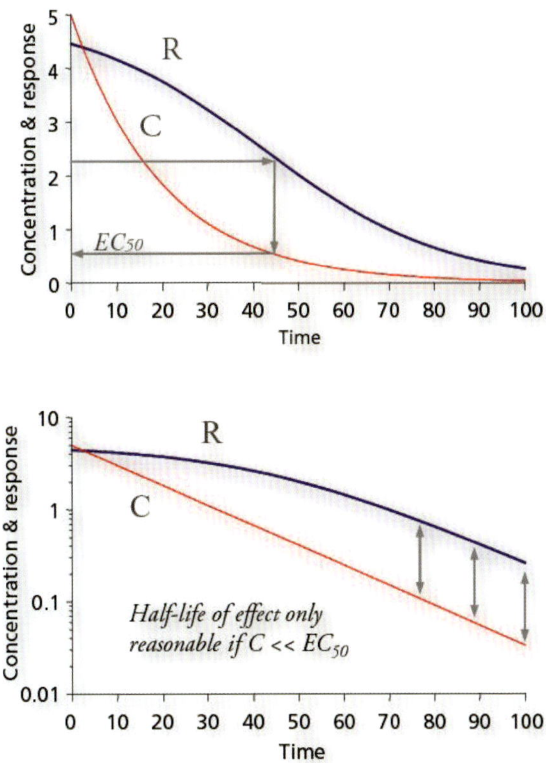

Figure 5.6
Concentration-time and response-time in a linear scale diagram (upper plot) and semi-logarithmic diagram (bottom plot). Potency EC_{50} denotes the plasma concentration at half maximal drug-induced response. Note that concentration and response are plotted on the same scale (Y-axis). Note that the plasma concentration and pharmacological response fall in parallel (indicated by three vertical double arrows) at concentrations far below the EC_{50} value provided instantaneous equilibrium and that the response does not approach a baseline value asymptotically.

When does a change in the plasma half-life impact the duration of response? Two typical situations that one may encounter clinically are shown in Figure 5.7. Imagine that the target plasma concentration falls below the potency EC_{50} concentration of the pharmacological response. That means that the response-time course will mimick the plasma concentration to a great extent. The half-life of response will be similar to the drug plasma half-life according to the picture of Figure 5.7 (right). The response will fluctuate in a similar manner as the plasma concentration (red curve in right hand plot). On the other hand, if the plasma concentration exceeds the EC_{50} concentration, indicated by the dashed blue line, then the pharmacological response (blue curve in the right hand plot) will be independent of the plasma concentration (blue curve) during a large extent of the dosing interval. In this case, the impact of a change in the plasma half-life of test compound is less important because plasma drug concentration will exceed the EC_{50} value to a large extent of the dosing interval. It will 'stay' at the flat portion of the concentration-response curve for a substantial period of time.

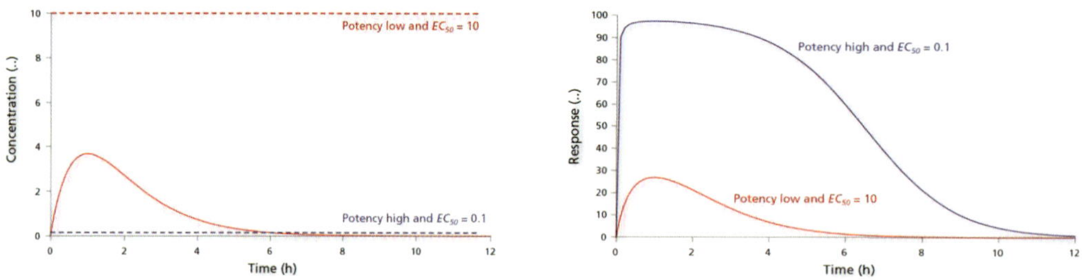

Figure 5.7
Schematic illustration of the concentration-time (left) and response-time (right) relationships for a compound assuming either high (EC_{50} 0.1) or low potency (EC_{50} 10). The right hand plots show the corresponding response-time courses.

To appreciate how the exposure to drug affects the response after chronic dosing, a simulation was done (Figure 5.8). The *pharmacodynamic* steady-state will, however, be highly dependent on the extent of exposure relative to the potency EC_{50} of the drug. If exposure is much less than EC_{50}, the pharmacological response will mimick the concentration-time course (Figure 5.8, upper part). On the other hand, if exposure substantially exceeds EC_{50}, pharmacodynamic steady-state may be established more quickly (Figure 5.8, bottom part). In the latter example, full response is established more or less instantaneously since the plasma concentration exceeds the EC_{50} value after the first dose.

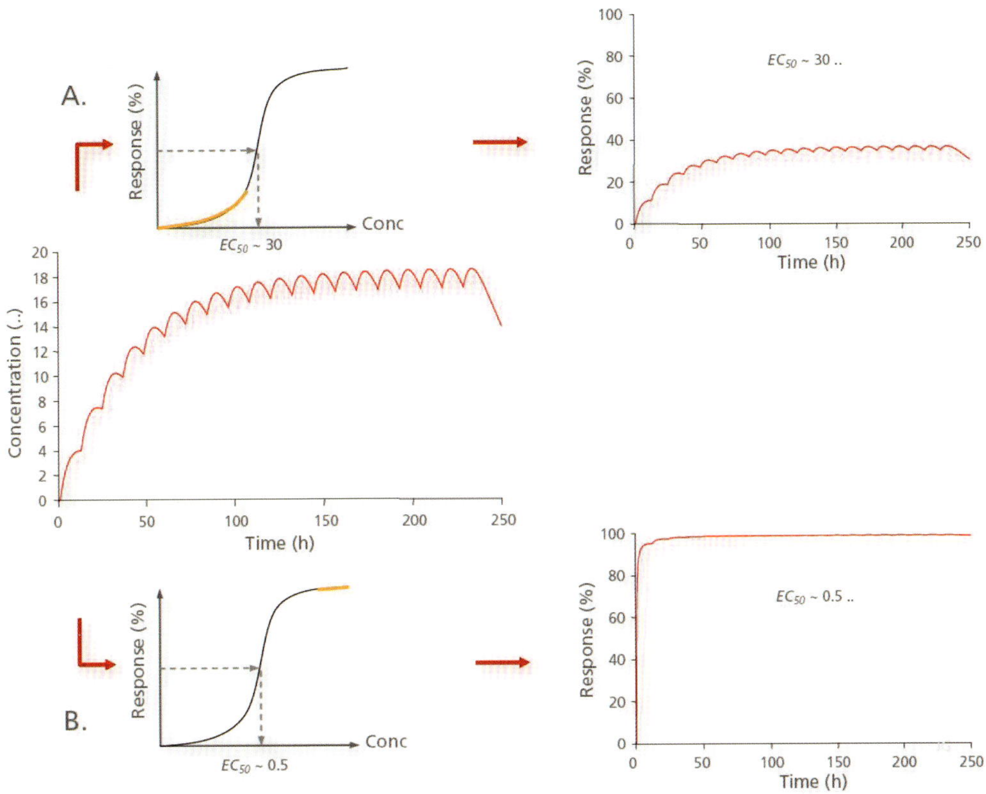

Figure 5.8
Illustration of the impact of different potency (EC_{50} values: 0.5 or 10) on the time to pharmacodynamic steady state and the actual steady-state response level for a system with instantaneous equilibrium between plasma concentration and a pharmacological response (left-hand graphs). **Case A:** Note how the time to pharmacodynamic steady-state mimicks the concentration-time course when exposure is much less than potency (upper right). **Case B:** When exposure exceeds the potency, pharmacodynamic steady-state is more or less instantaneous (lower right). The yellow intervals on the two concentration-response graphs indicate the apparent concentration range vis-á-vis the potency values.

The same reasoning is also valid for a multi-compartment pharmacokinetic system (Figure 5.9). Two simulations are done for a compound with pronounced multi-compartment properties. The concentration-time profile of an instant release *IR* tablet (blue curve, Figure 5.9) displays a bi-exponential decline post-peak. The kinetics of the same compound given as an extended release *ER* tablet was also simulated for comparison (red curve).

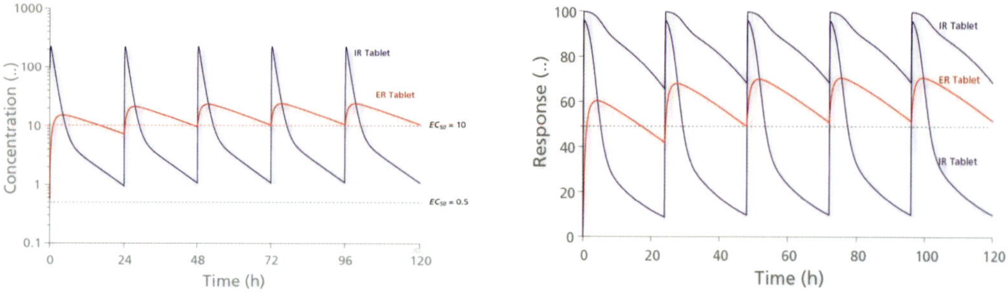

Figure 5.9
Left: Semi-logarithmic plot of simulated concentration-time data of an instant release IR tablet (blue curve) and extended release ER tablet (red curve). Two different target concentrations were aimed for (0.5 and 10 units). The 50 % of maximal drug induced response is illustrated by the horizontal dashed grey lines of 0.5 and 10 concentration units, respectively. The daily dose of the ER and IR formulations are the same. **Right**: Plot of simulated response-time data of an instant release IR tablet (blue curves) and extended release ER tablet (red curve). This aimed for two different plasma target concentrations (0.5 and 10 units). The 50 % of maximal drug induced response is illustrated by the horizontal dashed grey line. The upper blue line corresponds to an EC_{50} value of 0.5 and the lower blue line to an EC_{50} of 10 concentration units. The red simulated line uses an EC_{50} value of 10 concentration units.

When the EC_{50} is 0.5 units, the concentration-time course of the instant release *IR* tablet will always result in a response-time course which markedly deviates from the concentration-time course (Figure 5.9, *IR* tablet, upper blue curve). If, for example, an EC_{50} value of 10 concentration units is targeted, the response-time and concentration-time courses of the *IR* formulation will mimick each other (Figure 5.9, *IR* formulation lower blue curve). This results in a much shorter effect duration because the plasma concentration only stays above the EC_{50} for a brief time period. Large fluctuations will be observed in both concentration- and response-time courses. However, this can now be solved by means of an extended release *ER* formulation with a reduced peak-to-trough fluctuation and substantially extended duration of response (Figure 5.9, *ER* formulation red curve).

The conclusions to be drawn from this are that the plasma drug half-life only influences the response-time course when concentrations are below the EC_{50} for rapidly equilibrating concentration-response systems. In other words, a short plasma half-life may not necessarily imply a short duration of pharmacological effect of a potent drug, because plasma concentrations above the EC_{50} can be targeted for the major portion of the dosing interval. This means that whenever plasma half-lives are considered, exposures should also be related to the actual concentration range for effect and the potency of the compound. Delayed concentration-response relationships will be discussed in Chapter 6.

A gallery of different types of concentration-response relationships are shown in Figure 5.10. The linear concentration-response relationship is, of course, applicable at limited sections throughout the complete concentration-response (S-shaped) relationship. The log-linear and exponential concentration-response models cover a slightly larger interval than the linear model. The Hill-equation is generally applicable throughout the complete concentration-response curve, provided that data are symmetrical on both sides of the EC_{50} value on an effect *versus* log-concentration scale. Note that Figure 5.10 shows positive (stimulatory) responses. The same class of models could equally well be applied for inhibitory responses, for example, when the pharmacological response decreases from a baseline effect E_0 with increasing drug plasma concentrations.

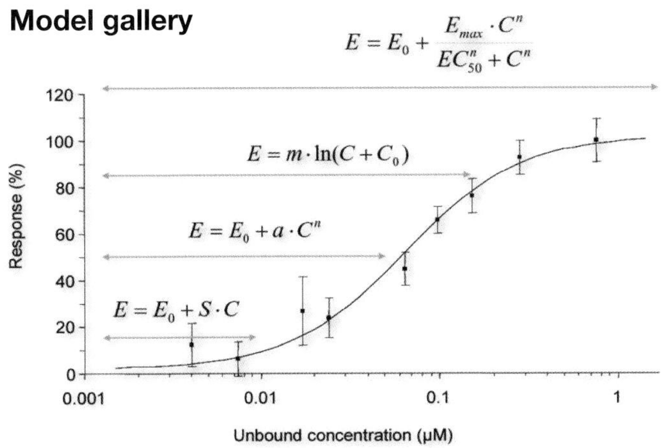

Figure 5.10
Experimental concentration-response data (filled circles, and their standard deviation, vertical bars) superimposed on a fitted Hill-equation (solid line). The linear (bottom), exponential (above linear model), logarithmic (above logarithmic model) and the Hill (top) concentration-response relationships.

The utility of the *linear* concentration-response relationship will be demonstrated in Section 5.4, the *exponential* concentration-response relationship in Section 5.5, the *log-line*ar concentration-response relationship in Section 5.6 and the *Hill-equation* in Section 5.6. The Michaelis-Menten type of *binding* occupancy will be examined in Section 5.7.

5.4 Linear concentration-response relationship - case example

The Case Study to be examined highlights how response-time and concentration-response data are analyzed after acute oral dosing of 100 μg of an agonist (*Mw* 318) to a male dog. The first step is plotting the concentration-time, response-time, baseline (vehicle control) response-time and concentration-response data (Figure 5.11 and Table 5.2).

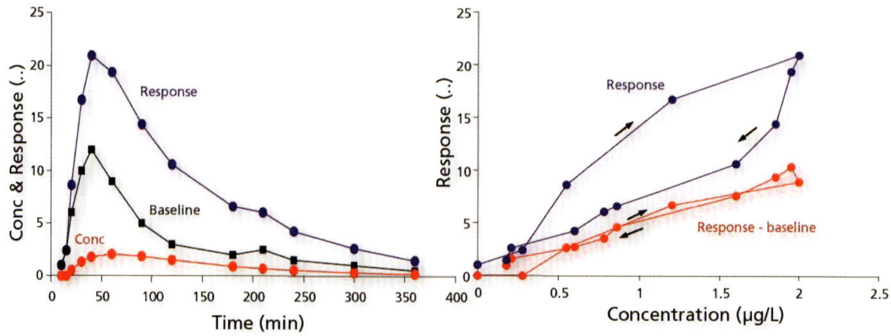

Figure 5.11
Concentration-time, baseline-(vehicle, placebo) time, and response-time data after an oral 100 μg dose of compound X (left). Right hand plot shows concentration-response data of total response (blue data) and baseline-subtracted response (red data). Note the clear hysteresis of the original data, which disappears for the baseline-subtracted data; instead a clear linear concentration-response relationship emerges.

The exploratory data analysis in Figure 5.11 shows that without baseline correction, a clear clock-wise concentration-response relationship or hysteresis is observed (Figure 5.11, right hand plot, blue curve). When the large baseline drift is subtracted from observed response a straight linear relationship between plasma concentration and response appears (Figures 5.11 (right) and 5.12 (left)). It was also observed that a large portion of the observed response contained the time-dependent change in the baseline response. This drift in the baseline over time, confounded the interpretation of the original concentration-response relationship. However, by subtracting the baseline response from the total observed response after test compound treatment, a clearer picture of the true acute drug induced effect is detected (Figure 5.12, Equation (5:1)

$$R_{obs} - BL(t) = E \tag{5:1}$$

where R_{obs}, $BL(t)$, and E are the overall test compound measured response, baseline or placebo response, and the pure test compound derived response, respectively. Since the true drug-induced response shows a simple linear relationship without an intercept of the response axis, we propose the linear concentration-response model in Equation 5:2 to be fitted to the baseline subtracted concentration-response data

$$E = a \cdot C \tag{5:2}$$

where a is the slope parameter that relates test compound concentration C to the pharmacological response E. An initial estimate of a is obtained by means of Equation 5:3

$$a = \frac{R_1 - R_2}{C_1 - C_2} = \frac{10.4 - 0.9}{1.95 - 0.18} \approx 5.4\, responseunits \cdot conc^{-1} \tag{5:3}$$

where a is estimated to be 4.94 ± 3 CV % when fitting Equation 5:2 to the baseline corrected concentration-response data.

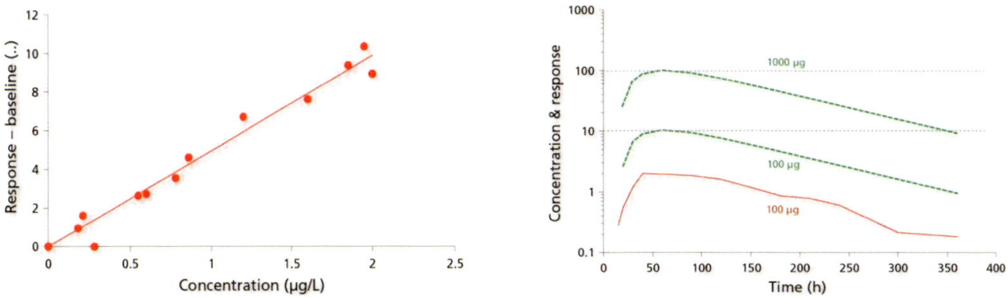

When the agonist was given daily as an oral dose of 100 µg over a period of 3 months to a group of male dogs, a different drug induced concentration- response relationship was seen with the biomarker response (Figure 5.13, grey dashed line). The same average concentration gave a 50–60 % lower response in biomarker compared to acute dosing. However, the disease marker response, which serves as the primary clinical endpoint appeared to be unaffected by tolerance. This shows a discrepancy between the clinical endpoint (disease marker) and what was believed to be a surrogate, namely the biomarker. The biomarker may still be used for screening purposes of target engagement. The disease model response can only be obtained after a 3–4 month dosing period at pharmacodynamic steady-state. Such studies are very costly, time- and resource-demanding (and are frequently performed in contract research laboratories).

Figure 5.12
The left hand plot shows concentration (µg $\cdot L^{-1}$) and baseline subtracted response (biomarker response) of the agonist. Red filled circles are experimental data and the solid red line is the model (Equation 5:2) fitted concentration-response relationship. One µg$\cdot L^{-1}$ corresponds to 3.14 nM. The right hand plot shows observed plasma concentrations (red curve) after a 100 µg oral dose and the corresponding predicted response-time course (middle dashed green curve). The predicted response-time course after a 1000 µg oral dose is illustrated by the upper dashed (green) curve.

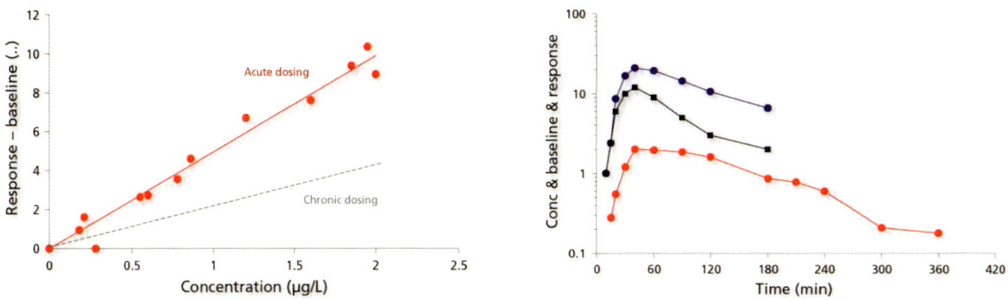

Figure 5.13
The left hand figure shows concentration (µg$\cdot L^{-1}$) and baseline subtracted response (biomarker) of the agonist after an acute dose (red filled circles and red solid line) of 100 µg and after 3 months daily dosing (grey dashed line) of 100 µg as an oral solution to a male dog.

The right hand figure shows total response-time data (blue), baseline response-time data (black) and baseline-subtracted concentration-time data (red), after acute dosing of 100 µg. The baseline and total response data are truncated after 180 minutes to more clearly show their differences in terminal slopes (Table 5.2).

How does one handle this new situation where there is a functional adaptation in the biomarker, which scales across different species, and where the disease model response seems to be unaffected? What kind of *in vitro* - *in vivo* biomarker correlation, and *in vivo* biomarker – *in vivo* disease marker correlation should be suggested? First the relationship between *in vitro* binding of test compound to the *in vivo* biomarker response after chronic dosing has to be established. Since both biomarker and disease marker are measured in the chronic disease model studies, one may plot *in vitro* concentration-binding, *in vivo* concentration biomarker and *in vivo* concentration disease model disease marker in the same graph. Using this graph the full relationship (*in vitro* binding to *in vivo* disease marker) is obtained. The *in vitro* binding will then guide the selection of more potent and selective compounds. These relationships with *in vitro* (hERG) and *in vivo* safety data (*Irwin, QT*) can then be updated in order to build a picture of the safety margin and selectivity at different plasma concentrations. The data shown in Table 5.2 below, correspond to graphed curves in Figures 5.11–5.13.

Table 5.2
Time (min), plasma concentration (μg·L^{-1}), model predicted plasma concentration (μg·L^{-1}), baseline response (..), test compound response (..) and baseline-subtracted response (..)

Time (min)	Conc (μg·L^{-1})	Pred conc (μg·L^{-1})	Baseline response (..)	Response (..)	Response – Baseline (..)
10	-	-	1	1	0
15	0.28	-	2.4	2.4	0
20	0.55	0.53	6	8.6	2.6
30	1.2	1.34	10	16.7	6.7
40	2	1.79	12	20.9	8.9
60	1.95	2.07	9	19.4	10.4
90	1.85	1.88	5	14.4	9.4
120	1.6	1.52	3	10.6	7.6
180	0.86	0.92	2	6.6	4.6
210	0.78	0.71	2.5	6.0	3.5
240	0.6	0.55	1.5	4.2	2.7
300	0.21	0.32	1	2.6	1.6
360	0.18	0.19	0.5	1.4	0.9

5.5 Exponential concentration-response relationship

This Case Study demonstrates how concentration-effect data can be analyzed by means of an *exponential* concentration-response model. The data are not complete and may not suffice for a more in depth analysis. However, it is a real life application of the exponential concentration-effect relationship. The ambition is to identify similarities and differences between the two groups of animals, and then to incorporate this information into a model that simultaneously fits both datasets. As a result of using simultaneous regression, we would also redefine the model and propose a new study design.

A new experimental compound is being developed. Data on the plasma concentration and clotting time of compound X in normal and *vitamin K* deficient rats are shown in Figure 5.14 and Table 5.3.

Figure 5.14
Left: Response versus concentration of normal and *vitamin K* deficient rats (left). Solid lines denote Day 5 and dashed lines Day 10 of dosing. **Right**: Observed (filled red circles and filled black squares) and model (Equations 5:4-5:5) fitted (solid lines) response.

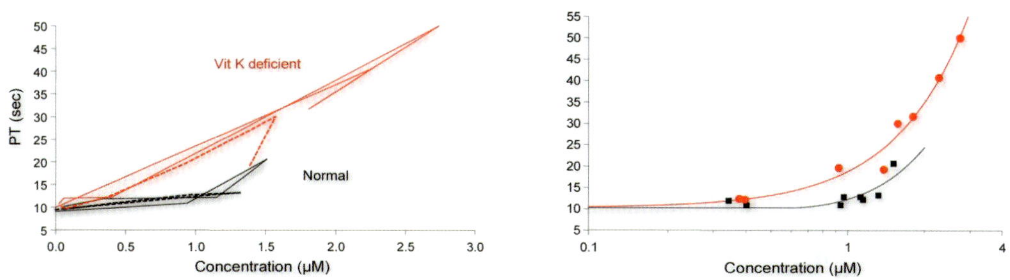

The pharmacodynamics of the two groups of animals starts deviating at plasma concentrations greater than 0.5 μM. Figure 5.14 (right) suggests an exponential relationship between plasma concentration C and pharmacological response E, and, therefore, the following general mathematical description of the data is suggested

$$E = E_0 + a \cdot C^b \tag{5:4}$$

The baseline parameter E_0 is the intercept on the response axis when C is equal to 0. An exponential term b is used because there seems to be a slight curvature in data when plotted on a log$C$ axis. The a parameter is just a proportionality factor of C. The relationship between response and concentration of normal animals is shown in Equation 5:5

$$E = E_0 + a \cdot (C - shift)^b \tag{5:5}$$

The *shift* parameter is the concentration difference between the diseased and normal animals at a response time of about 20 seconds, after which they start to run in parallel (See regression curves in Figure 5.14, right). A model is proposed that estimates the concentration that gives a 20 second coagulation response for each group of animals as secondary parameters. The equation for estimation of the concentration corresponding to a 20 second response is obtained by inserting the final parameter estimates of Equation 5:6 and solve for C_{e20}. In *Vitamin K* deficient animals C_{e20} becomes

$$C_{e20} = \left(\frac{20 - E_0}{a} \right)^{\frac{1}{b}} \tag{5:6}$$

and in normal animals

$$C_{e20} = \left(\frac{20 - E_0}{a} \right)^{\frac{1}{b}} + shift \tag{5:7}$$

The reasons for fitting the two concentration-response datasets simultaneously are that they overlap and have two or more parameters in common. Results from simultaneously fitting the two datasets are shown in Figure 5.15 and Table 5.3. The reason why an exponential model is used is because that model best mimicks the datasets within the presented concentration-response range. Ideally, one may also want to fit a pharmacodynamic model based on the logarithm of plasma concentration. An E_{max}-type of model will, of course, mimick the data, but the parameters will be highly correlated and estimated with poor precision.

Figure 5.15
Response versus concentration of normal and *vitamin K* deficient rats. All data pooled for the respective animal groups. Linear concentration scale left and logarithmic right.

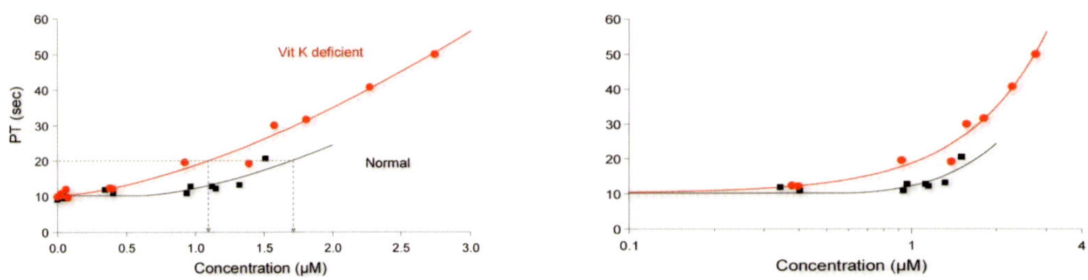

Parameter	Initial estimate	Final estimate	CV %
E_0	10	10.2	5
shift	0.5	0.61	16
a	10	8.5	11
b	1	1.5	8

Table 5.3
Initial and final estimates and CV % of the simultaneously fitting model.

A separate goal of this analysis was to estimate the plasma concentration giving a 20 second change in response for *Vitamin K* deficient and normal animals (Equations 5:4, 5:5 and Table 5.4). These values were 1.1 and 1.7 µM, respectively, which means that *Vitamin K* may be an important co-factor that has a direct bearing on parameters like the C_{e20} value. *Vitamin K* concentration in animals should therefore be incorporated into future studies.

Parameter	Final estimate	CV %
Vitamin K deficient C_{e20}	1.1	6
Normal animals C_{e20}	1.7	5

Table 5.4
The plasma concentration C_{e20} giving a 20 sec change in response

The present analysis is based on a naive pooled data approach and is lacking *Vitamin K* levels as an important co-variate. Consequently, the recommendation for future studies is a refined design of a bridging study based on steady-state data (day 14) including the time courses of test compound, response and *Vitamin K*. The experimental data are shown in Table 5.5.

Table 5.5
Time (h), total concentrations (µM) and effect data after oral doses of compound X 5 µmol·kg^{-1} to rats for 10 days

Group	Day	Time (h)	Conc (µM)	Effect (sec)	Group	Day	Time (h)	Conc (µM)	Effect (sec)
Normal	5	0	0	9.1	Diseased	5	0	0	9.97
		0.25	0.94	10.9			0.25	2.27	40.7
		0.5	1.51	20.6			0.5	1.8	31. 7
		1	1.14	12.2			1	2.74	50
		2.5	0.34	11.9			2.5	0.40	12.15
		5	0.044	10.1			5	0.06	12.0
		8	-	10.9			8	0.026	10.8
		24	-	9.7			24	-	10.7
	10	0	0	9.5		10	0	-	10.5
		0.25	1.12	12.8			0.25	1.39	19.2
		0.5	1.32	13.2			0.5	1.57	30.
		1	0.97	12.8			1	0.92	19.6
		2.5	0.4	10.9			2.5	0.38	12.3
		5	0.043	9.5			5	0.076	9.73
		8	-	9.6			8	0.026	10.3
		24	-	8.8			24	-	9.53

5.6 Hill *versus* log-linear models

In this Case Study the aim was to characterize the exposure-response (glucose lowering effect) of a test compound in female dogs after a single *vs.* repeated oral doses. A related aim was to assess rate of equilibrium between the plasma concentration (exposure) of the test compound and the glucose lowering effect (response). In this Section we show the:

- Design of a repeated dose regimen (in dogs)
- Exploratory (graphical) analysis of results
- Output from fitting two pharmacodynamic models to data
- Benefits from modeling data vis-á-vis hypothesis testing
- Simulated response-time courses of all animals and doses
- Goodness of fit criteria useful for model selection.

Three female dogs were given the test compound as oral solution once daily for 5 days. They were given either no drug (vehicle group, first week), 10 μmol·kg^{-1} (low dose, second study period) or 45 μmol·kg^{-1} (high dose, third study period) with a wash-out period of 9 days between each dose (Figure 5.16). The dogs were fed one hour before dosing and had free access to water during the experiment. The plasma concentration of the test compound was measured on day 1 and day 5 of dosing. Blood samples for subsequent analysis of the plasma glucose concentration were drawn in a time interval between pre-dose and 24 hours after the dose on administered day 1 and day 5.

Figure 5.16
Study design of the 5 day repeated oral dosing of the glucose-lowering test compound. The first period is a baseline study with vehicle treated animals, followed by a second and third period with 10 and 45 μmol·kg^{-1}·day^{-1}. Sampling of the test compound and the biomarker (glucose) was done day 1 and 5 of each study period.

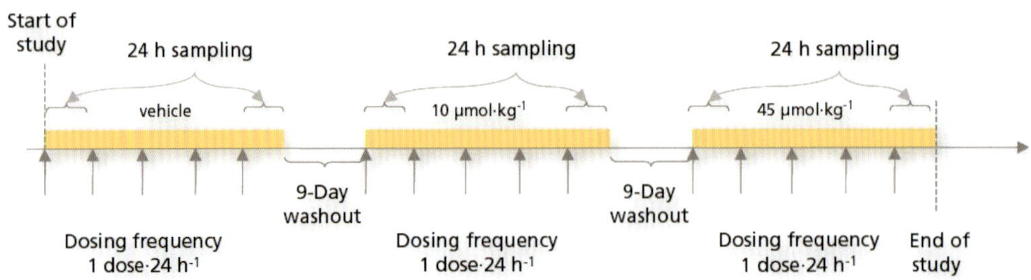

Figure 5.17 shows test compound concentration-time data of the 10 and 45 μmol·kg^{-1} doses on day 1 and day 5, respectively, in one dog. No accumulation was seen in test compound concentrations over the five day dosing period. There was a

slight tendency of a shift in t_{max} towards earlier time points when day 1 and day 5 data are compared. The corresponding glucose lowering effect in dog no 3 of the vehicle control, 10 and 45 μmol·kg-1 treatment arms, are shown in the bottom part of Figure 5.17. Each treatment arm shows almost superimposable data days 1 and 5.

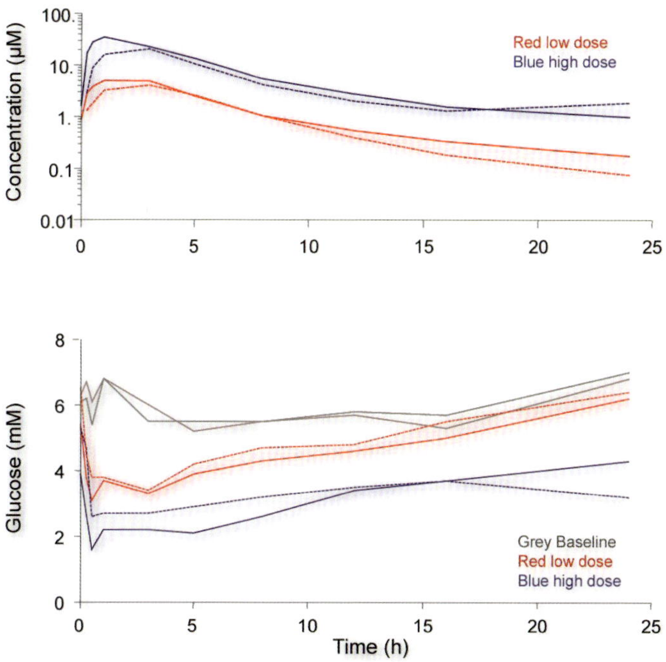

Figure 5.17
Upper figure shows a semi-logarithmic plot of concentration (μM) time (h) data of the test compound in plasma dosed with 10 (red lines) and 45 (blue lines) μmol·kg-1·day-1 in one on dog day 1 (dashed lines) and day 5 (solid lines) of dosing. The lower figure shows a plot of plasma glucose lowering (mM) time (h) data of test compound in plasma given the vehicle (control period, grey lines) and at 10 (red lines) and 45 (blue lines) μmol·kg-1·day-1 dose levels day 1 (dashed lines) and 5 (solid lines) of dosing.

When the concentration and glucose lowering response data of Figure 5.17 are combined into a concentration-response graph it can be seen that there is no hysteresis (Figure 5.18). This is a useful diagnostic of the lack of temporal differences between plasma concentration and drug response. Hysteresis means that the response is taking a different path relative to concentration when it rises from that taken when it falls. If hysteresis is observed it may be due to a distributional delay between plasma concentrations and the site of action, turnover mediated response, slow receptor-ligand on/off equilibrium, or active metabolite(s). The concept and consequences of hysteresis are discussed in greater detail in Chapter 6.

There is no clock-wise loop with either increasing and decreasing plasma drug concentrations day 1 or day 5. Rather, there is a random intertwined picture, which indicates a rapid equilibrium between plasma drug concentration and glucose lowering response.

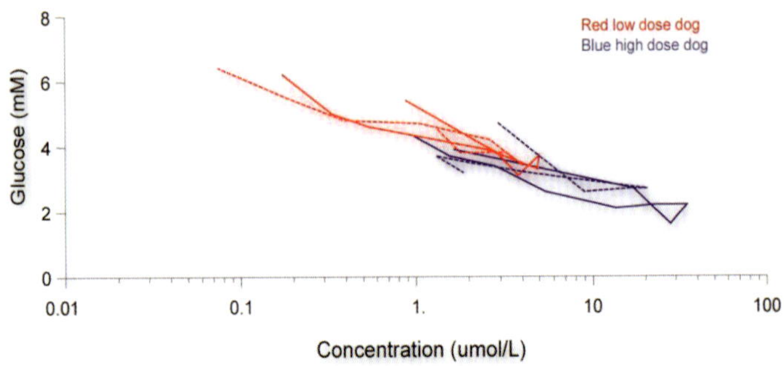

Figure 5.18
Semi-logarithmic plot of the concentration (μM) versus glucose lowering (mM) data of a test compound in plasma at 10 (red lines) and 45 (blue lines) $\mu molkg^{-1}day^{-1}$ dose levels in one dog day 1 (dashed lines) and 5 (solid lines) of dosing. There is no systematic trend in data when plotted in time-order – that is, no hysteresis.

The concentration-glucose response relationship is rather shallow and data from days 1 and 5 superimpose and display low within and between animal variability. In addition, no adaptation in the glucose response time courses were seen upon repeated administration (superimposable profiles days 1 and 5). No hysteresis or rebound was observed between the nadir and return to baseline due to the extended exposure to the test compound. Good coverage of glucose response was seen during most of the dosing interval at 10 $\mu mol \cdot kg^{-1}$. The lack of hysteresis allows the use of a rapid equilibrium approach since it can be hypothesized that either an inhibitory sigmoid I_{max} model, also called the inhibitory Hill equation, with a baseline parameter E_0, (Equation 5:8) or an inhibitory log-linear concentration-effect model (Equation 5:10) will accurately fit the data.

$$E = E_0 - \frac{I_{max} \cdot C^n}{IC_{50}^n + C^n} \tag{5:8}$$

where E_0, I_{max}, IC_{50} and n are the baseline, maximum inhibitory capacity of the test compound, potency and sigmoidicity factor, respectively. The plasma concentration C_{20} that corresponds to a specific effect E, say a 20 unit change from E_0 or E_{20}, becomes

$$C_{u,20} = IC_{50} \cdot \sqrt[\frac{1}{n}]{\frac{E_0 - E_{20}}{I_{max} + E_{20} - E_0}} \tag{5:9}$$

An alternative Equation to equation 5:8 is the inhibitory log-linear model.

$$E = E_0 - m \cdot ln(a \cdot C + 1) \qquad (5{:}10)$$

where E_0, m and a are the baseline value, slope of the log-linear function and a proportionality factor, respectively. When plasma concentration C is zero (0), Equation 5:10 is reduced to

$$E = E_0 - m \cdot ln(a \cdot 0 + 1) = E_0 - m \cdot ln(1) = E_0 - m \cdot 0 = E_0 \qquad (5{:}11)$$

Figure 5.19 shows schematically the behaviour of the inhibitory I_{max} model (Equation 5:8, left) and the log-linear equation (Equation 5:10, right).

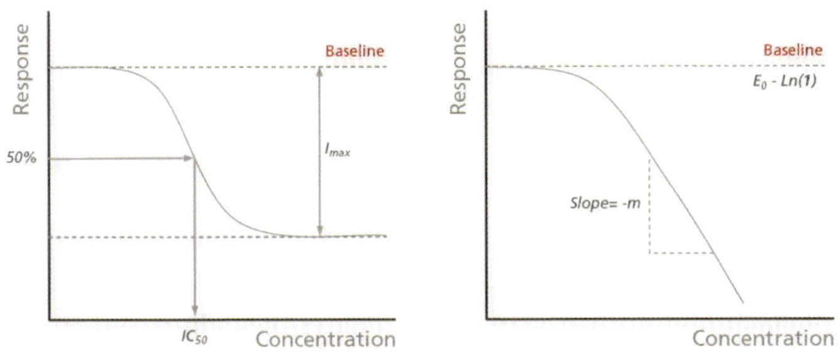

Figure 5.19
Left hand figure showing the concentration-response relationship of the inhibitory I_{max} model (Equation 5:8). Right hand figure showing the concentration-response relationship of the log-linear model (Equation 5:10).

The next step in the analysis is to obtain the initial estimates of the respective models parameters, fit the models to the data, and then analyze the output. If it is assumed the glucose baseline response E_0 is approximately 6 mM and the minimum response 2 mM, then the maximum reduction of response is 4 mM which corresponds to I_{max}. The plasma concentration corresponding to 50 % of that reduction, giving a glucose level of 4 mM (half-distance between 2 and 6 mM), is about 2 µM on the concentration axis, which corresponds to the potency IC_{50}. The initial estimate of the exponent n is assumed equal to unity. In Equation 5:12 the m parameter corresponds to the slope of the log-linear portion of the curve according to Figure 5.19.

$$slope = -m = \frac{E_1 - E_2}{ln(C_1 / C_2)} = \frac{6-2}{ln(0.1/30)} \approx -0.7 \qquad (5{:}12)$$

The a parameter Equation 5:10 is estimated by means of Equation 5:13 where one effect measure (*e.g.*, $E = 2$) is used together with the corresponding plasma concentration (30 µM)

$$2 = 6 - 0.7 \cdot ln(a \cdot 30 + 1) \tag{5:13}$$

which rearranged gives an a of approximately 10 µM^{-1}.
The question then arises which one of the *log-linear-* or inhibitory I_{max} models (Equation 5:8 or 5:10) fits the data best? An acceptable fit may not necessarily mean that the model is right. In light of the present experimental range and variability of data, both models fit the data equally well (Figure 5.20) with the sigmoid I_{max} model having a lower weighted residual sum of squares *WRSS* and higher parameter precision (Tables 5.6 and 5.7). Other goodness-of-fit criteria, such as the Akaike Information Criterion *AIC*, are basically the same for the two models. The condition number is far smaller (smaller is better) for the inhibitory I_{max} model than for the log-linear model, in spite of the fact that the former has two more parameters. The inhibitory I_{max} model also displays slightly more randomness in the residuals, in that the residuals change their signs more often.

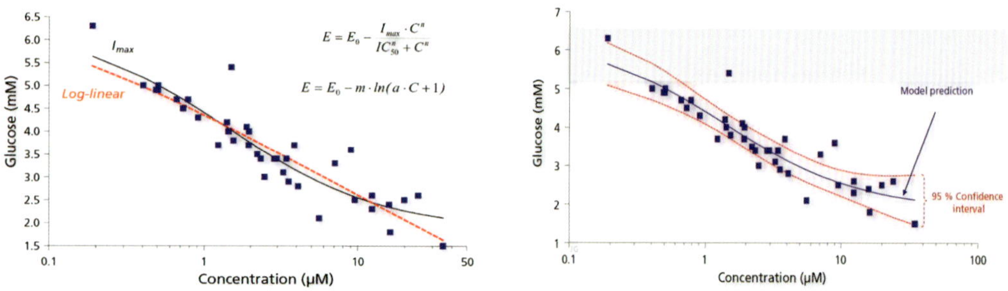

Figure 5.20
Left: Experimental (symbols) concentration-response data of one dog superimposed upon model predicted (solid line denotes the I_{max} model and dashed red line the log-linear model) of compound X. **Right**: Experimental data of the same dog (solid blue squares) superimposed on I_{max}-model predicted concentrations (solid blue line) and the 95 % confidence interval (dashed red lines) of the inhibitory I_{max} model (blue line).

It was possible to utilize experimental data by simultaneously fitting all plasma concentration-glucose lowering response data of day 5 with an I_{max}-model. Since baseline and low-dose data and low- and high-dose glucose data overlap, the same model and parameters could be used for the complete concentration-glucose lowering response course. Not only the point estimate of response for each plasma concentration was predicted but also the precision of the regression in terms of a confidence intervals. This demonstrates the real utility of regression modelling *vis-á-vis* hypothesis testing of changes in the average plasma glucose concentrations from the baseline.

Table 5.6
Final parameter estimates of fitting Equations 5:8 and 5:10 to data of three dogs

	I_{max}-model			Log-linear model		
Dog No	Parameter	Estimate	CV%	Parameter	Estimate	CV%
1	E_0 (mM)	6.1	5	E_0 (mM)	6.2	5
	I_{max} (mM)	4.4	16	m	0.7	10
	IC_{50}/IC_{u50} (μM)	1.9/0.5	36	a (μM^{-1})	9.7	65
	n	0.8	27			
2	E_0 (mM)	6.3	5	E_0 (mM)	6.3	5
	I_{max} (mM)	4.4	13	m	0.8	10
	IC_{50}/IC_{u50} (μM)	1.4/0.4	28	a (μM^{-1})	12.	62
	n	0.9	23			
3	E_0 (mM)	6.0	6	E_0 (mM)	6.2	5
	I_{max} (mM)	3.9	26	m	0.7	14
	IC_{50}/IC_{u50} (μM)	2.2/0.6	56	a (μM^{-1})	6.5	80
	n	0.8	36			

Combining a repeated dose regimen with the concentration-response relationship of the I_{max} model in Figures 5.19 and 5.20 and Equation 5:8 jointly with the pharmacokinetic concentration-time course of test compound X in plasma gives the repeated dose response-time courses of glucose after 5 doses at the low and high dose level in the experimental dog population (Figure 5.21).

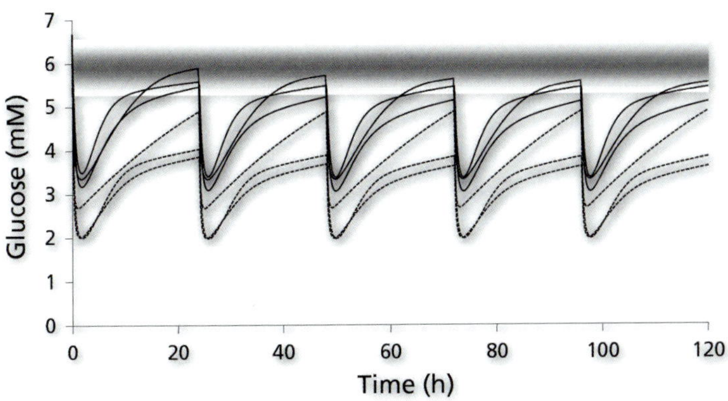

Figure 5.21
Simulated response (glucose, mM) in three dogs at the 10 (solid lines) and 45 (dashed lines) μmol · kg^{-1} · day^{-1} dose levels. The grey shaded horizontal area represents the baseline variability in control animals.

The pharmacodynamic conclusions to be drawn from this exercise are that if the target concentration falls below the IC_{50} value we will have a pharmacodynamic time profile mimicking the plasma concentration-time profile independently of a short or long plasma concentration half-life (Figure 5.22, yellow PD time courses). If the plasma target concentrations exceed the IC_{50} value, the pharmacodynamic time profile will display a longer duration than the plasma concentration-time profile, and be more difficult to predict due to its nonlinear nature.

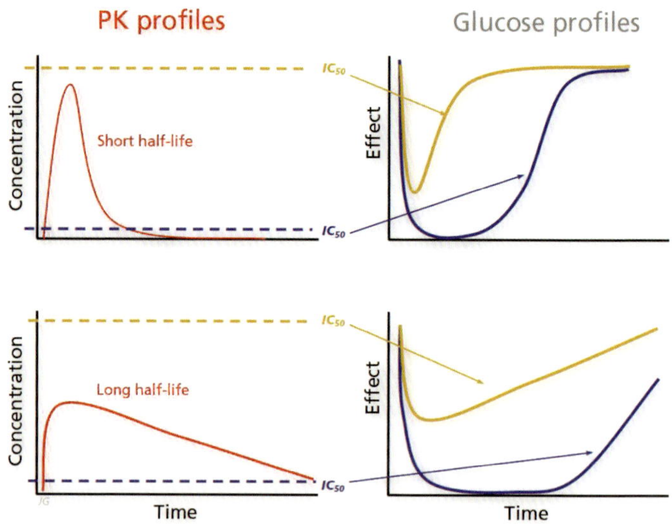

Figure 5.22
Schematic illustration of the concentration time course of rapid (upper left) and slow (lower left) kinetics. The horizontal dashed lines illustrate the low potency (upper) and high potency (upper) examples, respectively. The right hand figures show the corresponding effect time courses following rapid (upper right) and slow (lower right) kinetics at high (blue) and low (yellow) potency, respectively.

Table 5.7 shows the *Goodness-of-fit criteria* of both models for each animal dataset. The *Goodness-of-fit criteria* favour the inhibitory I_{max} model. There is one exception and that is a slightly lower *AIC* of the log-linear model fitting data of animal 1 and 3. For example, comparing the *AICs* of the I_{max} model (95.3) to that of the log-linear model (94.9) of dog no. 3, the latter model seems to be the model of choice in spite of the fact that *WRSS* are 8.87 and 9.22, respectively. This is a typical situation where the penalty function '$+ 2 \cdot n_{par}$' (Equation 5:14) takes its toll. The *Akaike Information Criterion AIC* is estimated as

$$AIC = n_{obs} \cdot \ln(WRSS) + 2 \cdot n_{par} \tag{5:14}$$

Where n_{obs}, $WRSS$ and n_{par} are the number of observations, the weighted residual sum of squares and the number of parameters, respectively.

Table 5.7
Goodness-of-fit criteria of the I_{max}- and log-linear models

	I_{max}-model		Log-linear model	
Dog No	**Parameter**	**Value**	**Parameter**	**Value**
1	*Condition no.*	15	*Condition no.*	315
	WRSS	7.39	*WRSS*	7.47
	+/-	22	+/-	20
	AIC	88.0	*AIC*	86.4
2	*Condition no.*	12	*Condition no.*	368
	WRSS	6.93	*WRSS*	7.61
	+/-	25	+/-	21
	AIC	85.46	*AIC*	87.19
3	*Condition no.*	22	*Condition no.*	194
	WRSS	8.87	*WRSS*	9.22
	+/-	17	+/-	15
	AIC	95.3	*AIC*	94.9

Even though plasma concentration monitoring is unlikely to be used in these circumstances, it is still important to delineate the determinants of the time course of response to improve the general understanding of the situation and also to optimize treatment procedures and drug use.

5.7 Michaelis-Menten type of binding model

This Case Study examines data on a new experimental compound that is being developed for a psychiatric disorder. Data on time, plasma concentration and binding of X to specific brain receptors are obtained from a study shown in Figure 5.23. The effective half-life of drug in the body is 2–3 hours.

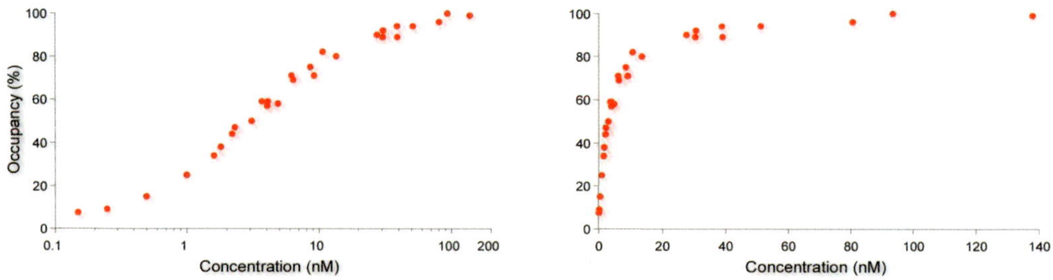

Figure 5.23
Semi-logarithmic scale of observed unbound brain concentration *versus* receptor occupancy (left) and linear scales (right) of the novel compound.

Part I: The dissociation constant K_d is obtained from the left-hand graph in Figure 5.23 at a concentration corresponding to 50 % occupancy which occurs at approximately 2–4 nM. A more precise value can be derived from fitting Equation 5:15 to the data (Table 5.9)

$$Occupancy(\%) = \frac{B_{max} \cdot C_u^n}{K_d^n + C_u^n} \tag{5:15}$$

Figure 5.24 shows the behavior of Equation 5:15. B_{max}, K_d and n denote the maximum binding concentration, placement (affinity) along the concentration axis, and curvature at the inflection point, respectively.

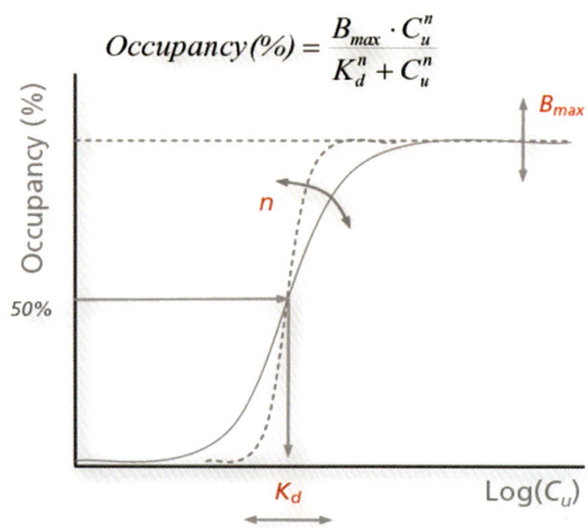

Figure 5.24
Schematic illustration of the behaviour of Equation 5:15. A change in B_{max} vertically shifts the top of the curve. The K_d parameter gives the location of the concentration-response relationship along the concentration axis. The n parameter changes the steepness of the curve at K_d.

The maximum binding concentration B_{max} parameter is normalized to 100 %. The corresponding half-life of k_{off} is

$$t_{1/2k_{off}} = \frac{ln\,2}{K_d \cdot k_{on}} \tag{5:16}$$

where K_d is the final estimate of the dissociation constant from the regression, and k_{on} is either 0.7 or 0.027 $h^{-1}{\cdot}C^{-1}$.

$$K_d = \frac{k_{off}}{k_{on}} = \begin{cases} k_{off} = 3 \cdot 0.7 = 2.1\,h^{-1} \Rightarrow t_{1/2k_{off}} = \dfrac{0.693}{2.1} \approx 0.3h \approx 20\,min \\ k_{off} = 3 \cdot 0.027 = 0.081\,h^{-1} \Rightarrow t_{1/2k_{off}} = \dfrac{0.693}{0.081} \approx 8h \end{cases} \tag{5:17}$$

The results from regressing fitting the data are shown in Figure 5.25 and Table 5.8.

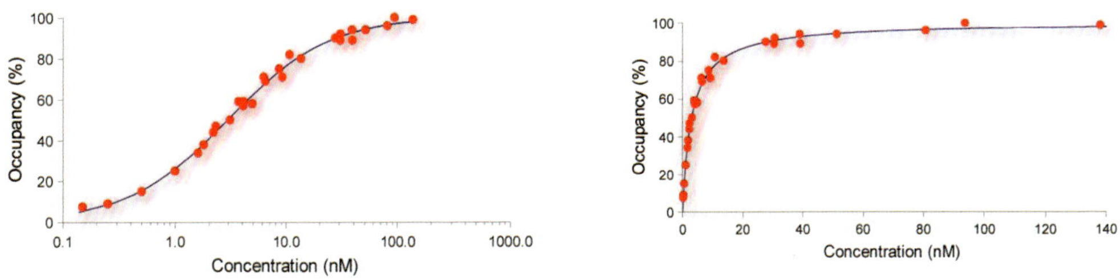

Figure 5.25
Observed (red symbols) superimposed on predicted (solid lines) occupancy *versus* concentration on semi-logarithmic (left) and linear scales (right).

The drug-receptor dissociation constant K_d is estimated to about 3 nM (CV % 4), the sigmoidicity factor n 0.95 (CV % of 4). For a drug-receptor dissociation constant K_d estimated to be 3 nM and k_{off} values of 0.27 and 0.081 $h^{-1}{\cdot}nM^{-1}$, the half-lives of k_{off} are 0.33 and 8 hours, respectively. This implies that a plasma concentration about 3 nM should be the target in at least one dose group. Based on the k_{off} values above, a time to pharmacodynamic steady state of approximately one hour (3–4 times 0.33 h) and 30 hours (3–4 times 8.7 h) may also be expected. This Case Study has shown how to apply a receptor binding model to real life in vivo data. The underlying assumption is that ligand binding to or dissociation from the receptor is not the rate limiting step. There is very little scatter in the data, also suggesting an uncomplicated relationship between unbound brain concentration and binding. The prediction of a k_{off} half-life of 8.7 h (k_{off} 0.081 h^{-1}) is not plausible, and neither is a k_{on} of 0.027 $h^{-1}C^{-1}$.

Table 5.8
Final model parameter estimates and their corresponding secondary parameters

Parameter	Primary/secondary	Final estimates	CV %
B_{max}	Primary model parameter	101	1
K_d (nM)		2.96	4
n		0.95	4
$T_{1/2koffA}$ (h)	Secondary parameter	0.33	4
$T_{1/2koffB}$ (h)		8.0	4

Table 5.9
Unbound brain concentration (nM) and receptor occupancy (%) data

Conc (nM)	Binding (%)	Conc (nM)	Binding (%)
0.25	9	8.6	75
0.15	7.5	6.4	69
1	25	6.2	71
0.5	15	3.7	59
4.9	58	13.6	80
1.8	38	4.1	59
9.2	71	93.6	100
4.05	57	30.4	89
2.31	47	138	99
1.6	34	27.6	90
3.1	50	80.7	96
2.2	44	39.1	89
38.9	94	51.2	94
10.7	82	30.6	92

A pharmacodynamic consideration in this context concerns the common assumption that there is a one-to-one relationship between receptor occupancy and biological response, and that EC_{50} values are suitable estimates of the desired level of occupancy relevant to therapeutic efficacy. Although this is frequently a very reasonable approach, there are notable deviations and aspects that may blur the picture – such as that the fractional target occupancy required for different ligands to exert biologically meaningful effects may be appreciably lower (*cf.* Figure 5.26) or higher than 50 %. As an illustration, direct dopamine D_2 receptor agonists may elicit marked biological responses occupying only small fractions of the corresponding receptor population – say, 2 % occupancy resulting in 50 % of maximum response in a cell system (*e.g.*, Burris *et al.*, 2002) – whereas traditional D_2 receptor-blocking drugs typically require 65–80 % occupancy for optimal antipsychotic efficacy in man (*e.g.*, Farde *et al.*, 1992). While receptor antagonist ligands in general display biologically relevant fractional occupancy in the upper range, the activity and behaviour of (in particular, partial) agonists may vary markedly between systems and experimental conditions. The apparent 'mismatch' between fractional occupancy and biological response readout may involve various different contributing factors, including, *i.a.*, the number of available receptors (B_{max}), level of endogenous transmitter tone, ligand on/off rate, efficiency of the signal/response transduction system, potential stoichometry and "promiscuity" in the latter (ability of a compound to elicit cellular responses via more than one receptor coupling cascade; *viz. e.g.*, G_s *versus* G_i-coupled pathways) in relation to the selectivity/specificity of the activating ligand (for further discussion the interested reader is referred to, *e.g.*, Kenakin, 2006, and references cited therein). Taken together, the possibility that different biological responses are elicited at varying degrees of fractional receptor occupancy needs to be factored in when establishing exposure-response relations, secondary pharmacology and safety margin ratio estimations, as well as dose predictions from preclinical data to man.

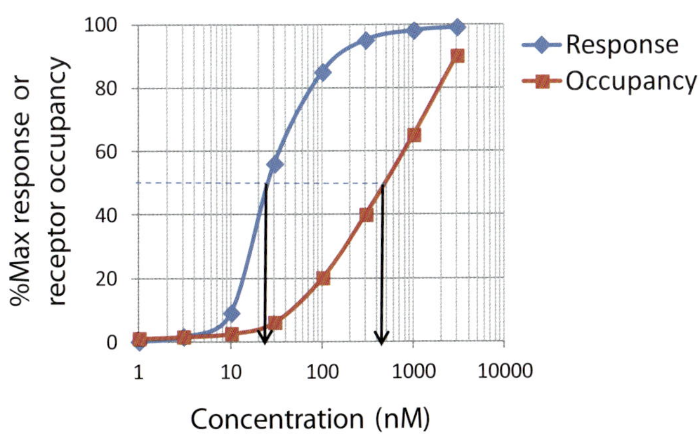

Figure 5.26
Schematic semi-logarithmic plot reflecting non-linear relations between exposure, biological response (blue) and receptor occupancy (red) - as illustrated, *e.g.*, by findings in Burris *et al.* (2002). Dashed line and arrows depict the corresponding apparent EC_{50} readouts; note the difference in steepness and the >10-fold difference between response and occupancy EC_{50}.

5.8 Case Study example: Analysis of brain occupancy and exposure data of several species

The aims of this Case Study are

- to analyze concentration and binding occupancy data after oral administration of an CNS active compound to marmoset monkeys, cats and humans
- estimate the drug-receptor dissociation constant K_d
- assess the human dose based on occupancy data and data from a preclinical disease model.

Outline: A new antagonist is being developed for a novel mechanism of action in the treatment of schizophrenia. Data on total plasma concentration and brain occupancy of compound X are obtained from three studies in different species (Figure 5.27).

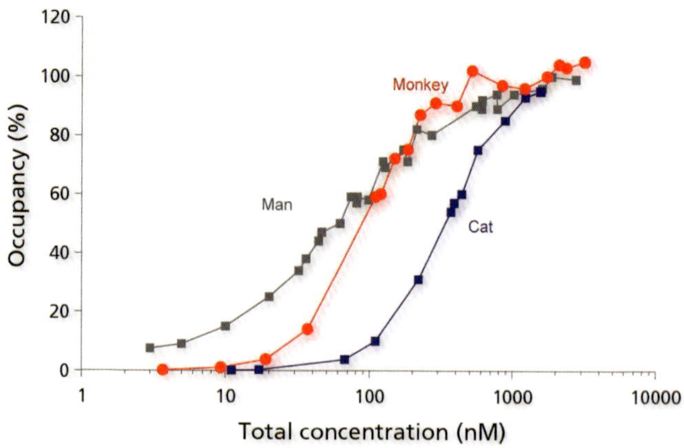

Figure 5.27 Semi-logarithmic scale of observed total plasma concentration *versus* occupancy data from marmoset monkey, cat and man. The curves are distinctly separated from each other because total concentrations are used.

Data from a separate *in vivo* disease model in marmosets showed a efficiency of 47 and 91 % at an average unbound plasma concentration of 2 and 40 nM, respectively. The plasma half-lives are 0.45 and 1.5 h in the cat and marmoset, respectively. The predicted plasma half-life in man is approximately 3 h based on allometric data in rat, cat, marmoset and beagle dog. *Practice example*: Contrast the plasma concentration half-lives with the predicted half-lives of k_{off} in available species and man. Analyze the data in Table 5.11 plasma concentration and brain receptor occupancy according to appropriate methods covered in Section 5.7. What conclusions about species similarities or differences, do you draw from this analysis? Why is K_d not enough for assessment of duration of response? What more parameter(s) is(are) needed?

Given a k_{on} of 0.7 and 0.027 $h^{-1} \cdot nM^{-1}$, respectively, what are approximately the half-lives $t_{1/2koff}$ of response and what are the limitations with these estimations?

Start plotting the data as shown in Figures 5.28.

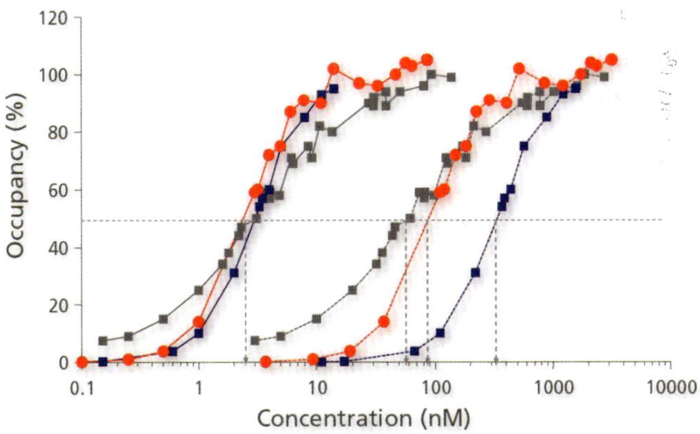

Figure 5.28
Semi-logarithmic scale of observed total (three right hand dashed lines) and unbound (left hand solid lines) plasma concentration *versus* occupancy data from marmoset monkey (red), cat (blue) and man (grey). Note how when unbound concentrations are used the three left curves nicely converge.

The pharmacography of Figure 5.26 shows the approximate dissociation constant K_d that is obtained from the left-hand graph at an unbound concentration C_u corresponding to 50 % occupancy. That occurs at approximately 2–3 nM. A more exact value coupled to its precision can be derived from fitting Equation 5:18 to the data.

$$Occupancy(\%) = \frac{B_{max} \cdot C_u^n}{K_d^n + C_u^n} \tag{5:18}$$

The unbound plasma concentration $C_{u,20}$ that corresponds to a specific occupancy, say 20 % or E_{20}, becomes.

$$C_{u,20} = K_d \cdot \sqrt[\frac{1}{n}]{\frac{E_{20}}{B_{max} - E_{20}}} \tag{5:19}$$

The maximum binding concentration B_{max} parameter is normalized to 100 %. The corresponding half-life of k_{off} is

$$t_{1/2k_{off}} = \frac{ln\,2}{K_d \cdot k_{on}} \tag{5:20}$$

The results from fitting the three datasets are shown in Figure 5.29 and Table 5.10.

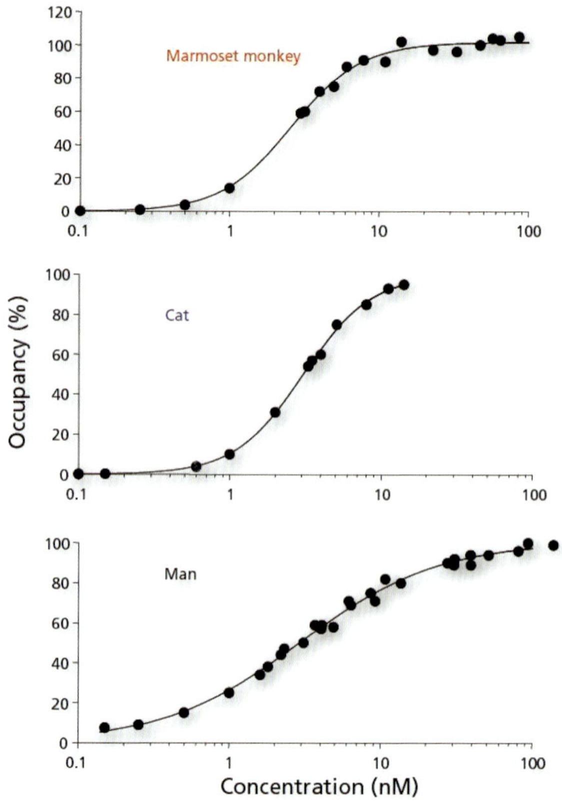

Figure 5.29
Semi-logarithmic scale of observed unbound (symbols) plasma concentration *versus* occupancy data from marmoset monkey, cat and man superimposed on fitted (solid lines) data. Note that marmoset and cat have steeper slopes than human occupancy data.

All parameters have high precision in all species (*CV* ranges between 1 to 7 %). The maximum binding concentration is normalized to about 100 %. Note the difference between the exponent *n* which is about 1 in human in comparisons 2 in marmosets and cats. The affinity between ligand and receptor is about 3 nM.

Table 5.10
Maximum binding concentration B_{max} (expressed as %), dissociation constant K_d (nM) and sigmoidicity factor n of compound X in mamroset monkey, cat and man.

	Monkey		Cat		Man	
Parameter	**Estimate**	**Precision (%)**	**Estimate**	**Precision (%)**	**Estimate**	**Precision (%)**
B_{max}	102	1	100	2	101	1
K_d	2.6	4	3.1	2	3	4
n	1.9	7	1.9	4	0.95	4

K_d is the estimate of the dissociation constant from fitting Equation 5:18 to data in Figure 5.29, and k_{on}, derived for monkey, cat and man are 0.11, 0.13 and 0.035 $h^{-1} \cdot C^{-1}$, respectively. This gives an opportunity to estimate for each species the first-order dissociation constants and their corresponding half-lives.

$$K_d = \frac{k_{off}}{k_{on}} \Rightarrow \begin{cases} k_{off} = 2.6 \cdot 0.11 = 0.29\, h^{-1} \Rightarrow t_{1/2k_{off}} = \dfrac{0.693}{0.29} \approx 2.4h & monkey \\ k_{off} = 3.1 \cdot 0.13 = 0.40\, h^{-1} \Rightarrow t_{1/2k_{off}} = \dfrac{0.693}{0.40} \approx 1.7h & cat \\ k_{off} = 3.0 \cdot 0.035 = 0.11\, h^{-1} \Rightarrow t_{1/2k_{off}} = \dfrac{0.693}{0.11} \approx 6.6h & man \end{cases} \tag{5:21}$$

The half-lives of k_{off} were 2.4, 1.7 and 6.6 h, which then should be compared to the plasma half-lives of 1.5, 0.45 and 3 h in marmoset monkey, cat and man, respectively. The longer half-life of k_{off} suggests a longer duration of action than could be predicted from the plasma half-life of 3 h *per se*. A half-life of k_{off} close to 7 h also suggests a convenient morning and evening dosing regimen of the drug provided safety allows wide fluctuations in peak-to-trough values. Data from a preclinical disease model are shown together with brain receptor occupancy data in Figure 5.30). The disease model efficacy is 43 and 91% at an unbound plasma concentration of 2 and 40 nM, respectively, are obtained from a four month chronic dosing study in marmosets. There is a striking consistency between animal and human occupancy data and the preclinical marmoset disease model. They all seem to operate within the same effective concentration range.

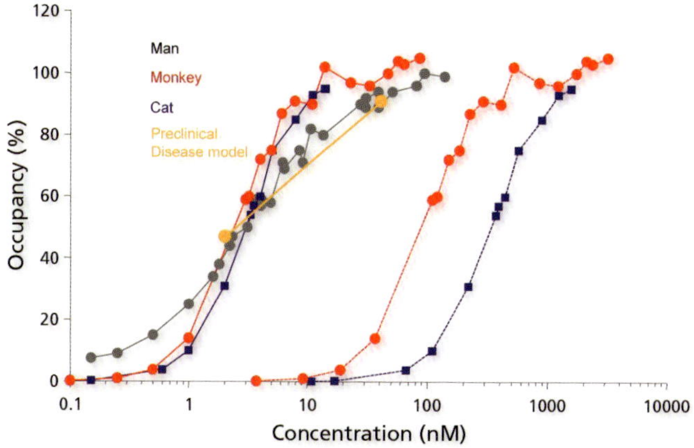

Figure 5.30
Semi-logarithmic scale of observed total (dashed lines) and unbound (solid lines) plasma concentration *versus* occupancy data from marmoset monkey, feline and man. Data from the marmoset disease model (yellow symbols and line) are superimposed. Note how nicely functional data superimpose on occupancy data when based on unbound concentrations.

Table 5.11
Total plasma concentration (nM) and receptor occupancy (%) data in marmoset monkey (f_u 2.7 %), Norwegian Blue feline (f_u 0.9 %) and man (f_u 5 %)

The consistency between the preclinical disease model and preclinical and clinical binding data is very convincing. The next step is then to take this compound into patients targeting the effective concentration range based on the binding-occupancy data.

Monkey		Cat		Man	
Conc (nM)	Occupancy (%)	Conc (nM)	Occupancy (%)	Conc (nM)	Occupancy (%)
3.70	0.16	11	0.1	3	7.5
9.30	1	17	0.25	5	9
19	3.8	67	3.8	10	15
37	14	110	10	20	25
110	59	220	31	32	34
120	60	370	54	36	38
150	72	390	57	44	44
185	75	440	60	46	47
226	87	570	75	62	50
290	91	890	85	74	59
407	90	1230	93	81	57
520	102	1570	95	82	59
850	97			98	58
1220	96			124	71
1740	100			128	69
2110	104			172	75
2370	103			184	71
3190	105			214	82
				272	80
				552	90
				608	89
				612	92
				778	94
				782	89
				1024	94
				1614	96
				1872	100
				2760	99

5.9 Summary and discussion of concentration-response relationships

This Chapter has focused on different concentration-response relationships where the equilibrium occurs rapidly between the two. No time delay (hysteresis) is seen between the concentration-time and the response-time courses. Table 5.12 summarizes some pivotal parameters related to nonlinear pharmacodynamics, how they are measured and how one apply them.

Table 5.12
Summary table of some pivotal/ pharmacodynamic parameters EC_{50}/IC_{50}, E_{max}/I_{max}, *n* and *selectivity*.

Parameter	What is it?	How is it obtained?	How do you apply it?
EC_{50}/ IC_{50}	Potency: The plasma concentration at 50 % of maximum drug induced response	By means of nonlinear regression of either concentration-response or response-time data	Guidance parameter for the effective concentration range
E_{max}/ I_{max}	Efficacy: The maximum drug induced response	By means of nonlinear regression of either concentration-response or response-time data	
n	Sigmoidicity factor	By means of nonlinear regression of either concentration-response or response-time data	If low it may indicate resistance to a specific compound. Be aware that *n* may change when you convert data from unbound to total concentrations since it is plasma protein binding dependent
Selectivity	Separation between primary and secondary effects	From the difference between the primary and secondary concentration-reponse curves	As a concentration-dependent measure of safety or selectivity margin in the comparison with other compounds

In various experimental settings responses to a drug may be ascribed being driven either by the peak exposure C_{max}, or by the drug exposure over a predefined time interval *AUC*. Needless to say, C_{max} and *AUC* parameters are intertwined and related to each other in many instances and it may not always be straightforward to discern whether a response is truly driven by one or the other, particularly when linear first-order kinetics is present.

Obviously, the actual shape of the exposure-time curve is instrumental in determining the expression of a response (be it therapeutically desirable or considered a safety/adverse event). The route and method of administration is thus of key importance to the manifestations of drug action. Administration modes circumventing barriers and first-pass metabolic degradation result in the highest and most rapid C_{max} peaks, as exemplified by parenteral bolus injection. A response that is strictly driven by the C_{max}, will vanish once drug elimination processes have taken the concentration below what was required for triggering (and sustaining) the response. For example, whereas for bactericidal aminoglycoside antibiotics the C_{max} to *MIC* (minimum inhibitory concentration) ratio appears to be the key driver of efficacy, other antibiotic drug classes mainly depend, albeit to differing extents, on *AUC/MIC*, or time *Cp*> *MIC*.

To shed some light on *AUC*-driven response, let us assume that a promising drug candidate has been discovered that displays excellent primary pharmacodynamic properties. However, while it is aimed at treating a condition that necessitates 24h coverage for optimal therapeutic effect, its half-life may be too short for this being achieved by means of once daily dosing. Even twice or thrice daily dosing may be inadequate because the administration of single high doses to sustain exposure above the therapeutic threshold across the 24h period results in excessive C_{max} peaks, and safety or tolerability issues as a consequence thereof. Still, in this very case it may be so that it is the *AUC* above the threshold concentration that determines the overall extent of drug action. The application of extended-release formulation approaches may provide a solution to the challenge, by allowing for administration of the very same daily drug load – calculated on an *AUC* basis – but in a flatter time-concentration fashion, thereby avoiding tolerability/safety issues while retaining the exposure level within desired therapeutic limits over the 24h-period.

The extent rather than the actual shape of the *AUC* impacts the pharmacological outcome, when the drug action is irreversible such as for omeprazole (active metabolite that binds irreversibly to the proton pumps in the gastric mucosa). It may also occur as if C_{max} rather than *AUC* is a better predictor of drug action when the drug action is reversible and the equilibrium time between plasma and the pharmacological response is rapid. Chapter 6 gives an introduction to real life Case Studies which exhibit temporal differences between plasma drug concentrations and the pharmacological response.

Chapter 6

Time Delays between Plasma Concentration and Response

6.1 What is this chapter about?

- This Chapter deals with temporal differences between concentration-time and response-time curves and the concept of *hysteresis*
- We will further discuss how to assess *potency, efficacy, steepness* and *selectivity* and factors determining the time delay, based on a combination of the concentration-time, response-time and concentration-response relationships
- The half-life of response will be examined from a pharmacodynamic perspective. For example, does a half-life of response *per se* really exist? This Chapter deals with how to quantify the half-life of response
- Finally, the pharmacodynamic model gallery is exemplified by three Case Studies using the *effect-compartment model,* the *turnover model* and the *receptor binding on/off model.*

6.2 Time delays observed as hysteresis

When instantaneous equilibrium occurs between the drug plasma concentration and the pharmacological response, we have a situation shown in Figure 6.1 (left). Every single plasma concentration corresponds to a single response independently if concentrations are ascending or descending. This is shown in the bottom left part of Figure 6.1. If there is a delay between the plasma concentration-time course and the pharmacological effect-time course, a situation arises that is illustrated by the right hand plots in Figure 6.1. The same plasma concentration occurring at two different time points results in two different responses. The response is lagging behind the plasma concentration both on the upswing and washout. There are temporal differences between the plasma concentration and response – equilibrium does not occur instantaneously.

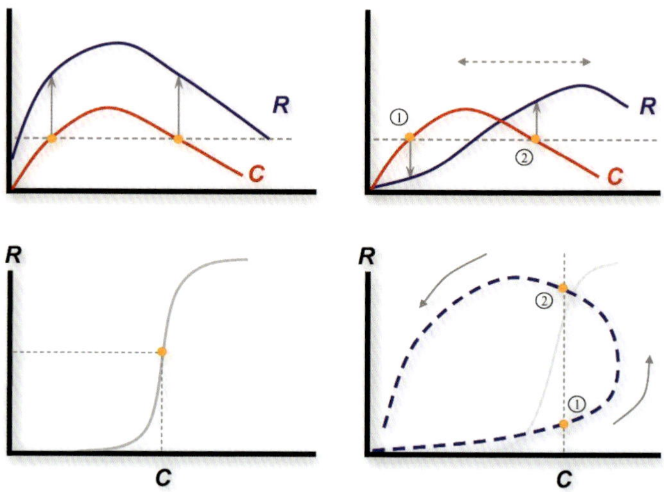

Figure 6.1
Left: Schematic illustration of rapid (direct) equilibrium between concentration (*C*) and effect (*R*) (upper and lower left figures).
Right: Delayed concentration-effect relationship (hysteresis, upper and lower right figures). The grey S-shaped line in the figure is the true concentration-response relationship at equilibrium.

Temporal divergencies like the aforementioned thus led to the erroneous conclusions that there was no relationship between the concentration of a drug in blood or plasma and the time course of response. However, as will be further discussed below, the pharmacological effects of many drugs lag behind the drug concentration in plasma – observed as *hysteresis* (Figures 6.1 right and 6.2). There is indeed a relationship between plasma exposure and pharmacological response, but it is a long-term one which will be more obvious at equilibrium. Such an example is the antianemic agent erythropoeitine (Epo) with a half-life of less than 10 hours. It reaches steady-state within two days although its therapeutic effect lasts for a month. The impact of Epo on the red blood cell count takes time to develop. Epo directly affects the production rate of red blood cells throughout its entire time course. The actual level of red blood cells is not only determined by production but also on the removal, and the lifespan of red blood cells (>100d) greatly outlasts the half-life of Epo. Thus, even a brief rise in Epo levels will result in a prolonged effect on red blood cell count.

Another example is the impact of cannabinoid receptor type 1 (CB1r) inverse agonists on body weight and fat content. CB1r inverse agonists decrease appetite and therefore in turn reduce body weight and fat stores. Effective plasma half-lives of these compounds typically range between 20 and 50 h, establishing steady-state within a week. However, the half-life of the adipocytes (fat cells) falls in the range of several weeks in experimental animals, and several counterbalancing adjustments to the CB1r inverse agonist-induced lowered food intake (and hence perceived "starvation state") are simultaneously triggered. Pharmacodynamic steady-state is established after months of dosing when rates of production and removal of adipocytes balance each other. The body weight reduction now fully reflects the increased removal of adipocytes. This apparent dissociation can be rationalized by the use of some of the approaches shown in Figure 6.3. The advent of the effect compartment, turnover and receptor occupancy models has given us tools to better explain and quantitatively tackle temporal differences. Using a model allows a more stringent design, analysis and interpretation of kinetic dynamic experiments. Using a model allows predictions to be made and the model further challenged with new experimental data.

The temporal differences between plasma concentration and response (Figure 6.1, right) may be due to distribution from plasma to biophase, active metabolites, initiation of changes of factors controlling the onset and offset of response (cf. *e.g.*, Epo and CB1r examples), or a slow on/offset of receptor occupancy. One of the primary goals in PKPD assessment is to identify the

Figure 6.2
Left hand column: Concentration-time and response-time courses for a system with rapid equilibria between concentration and response (upper) and the corresponding concentration-response relationship below.
Middle column: Concentration-time and response-time courses for a system with a slight delay between peak concentration and maximum response (upper) and the corresponding concentration-response relationship shown as a hysteresis loop below.
Right hand column: Concentration-time and response-time courses for a system with a large delay between peak concentration and maximum response (upper) and the corresponding concentration-response relationship shown as a hysteresis loop below. ① and ②, denote two time points where plasma concentration is the same. ③ denotes the peak concentration and its corresponding time. The weak solid grey lines represent the concentration-response profiles at equilibrium.

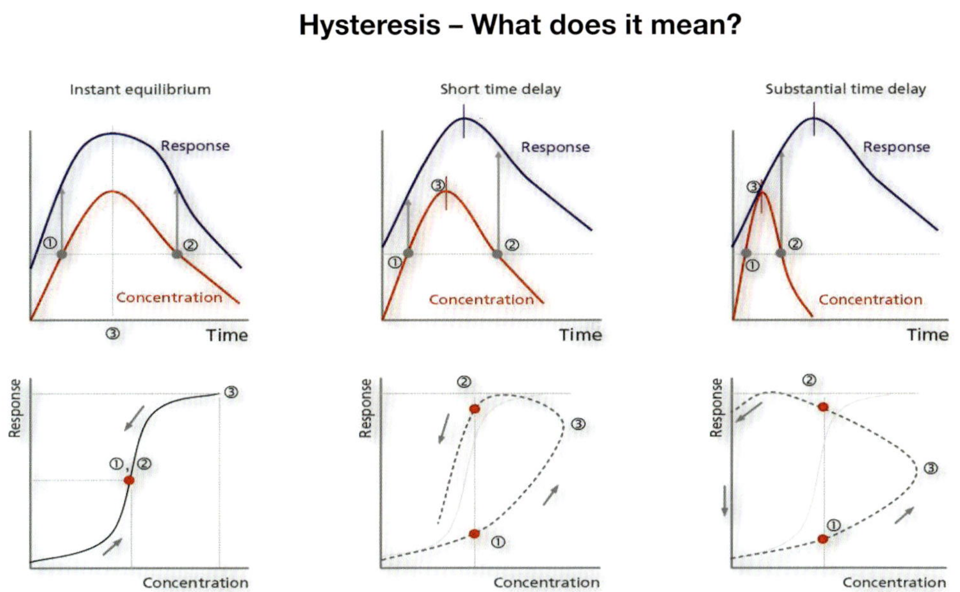

rate-limiting step along the concentration-response axis, and to judge whether this rate-limiting process is due to distributional, turnover or receptor on/off processes. When the system reacts more rapidly than distributional processes (schematically shown in Figure 6.3) then this is called distribution-rate limited dynamics and is discussed in Section 6.3.

Thus, the pharmacokinetic requirements for an optimal dynamic effect may vary substantially depending on the type of target and the position of the target in the biological system eliciting the effect. A generic way of working is therefore not possible and an integrated PKPD strategy has to be adopted using a case-by-case approach.

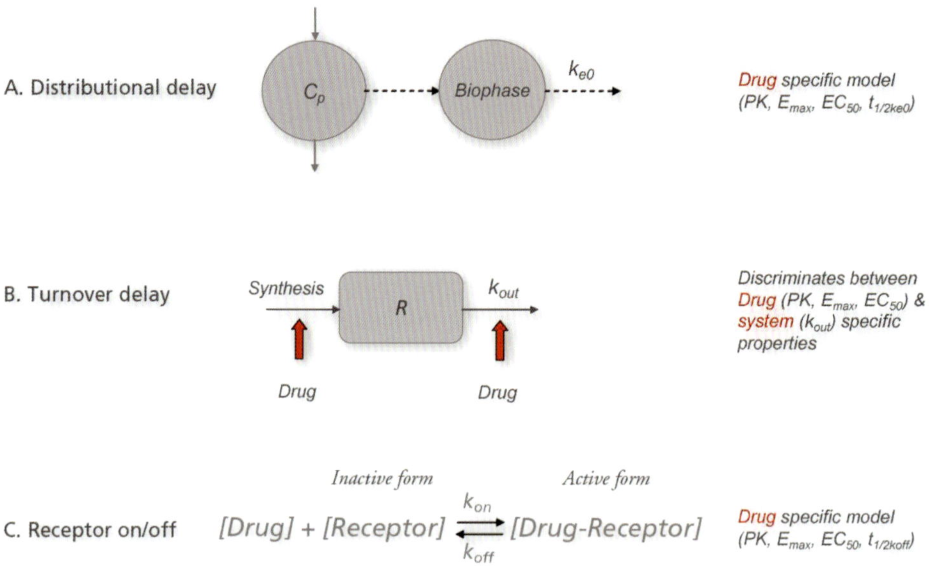

Figure 6.3
Schematic illustration of distributional models (A, effect compartment models), turnover models (B) and receptor on/off models (C). Note that distributional and receptor kinetic models do not discriminate between drug and system parameters. In turnover models there is a distinct differentiation between drug parameters (*E_{max}*, *EC_{50}*) and system properties (*k_{in}*, *k_{out}* etc.). In the case of turnover models, the system represents the physiology or pathology (disease) that we wish to model.

6.3 Delays caused by distributional processes

6.3.1 When are distributional models needed?

When a delay occurs between the plasma drug concentration and the pharmacological response it is observed as hysteresis in the concentration-response plot. When this delay is due to a *slow distributional process* between plasma and the effect site we apply an effect-compartment model (Figure 6.3 A.). The rate of change of the drug plasma concentration occurs more rapidly and peaks earlier than the pharmacological response-time course. The temporal differences between plasma concentration and response is resolved by inclusion of a hypothetical effect compartment. This requires an additional compartment between plasma and response. The time course of drug in the effect compartment peaks at the same time point as the pharmacological response (Figure 6.4) and is at equilibrium with the response at all times. The concentration within the effect compartment is also assumed to be equal to that in plasma at equilibrium (Figure 6.4).

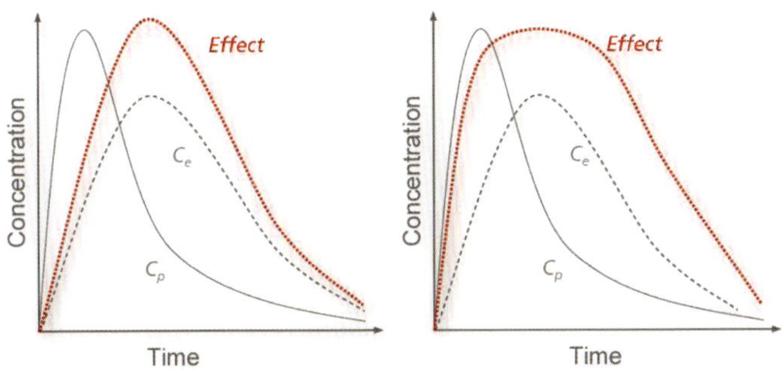

Figure 6.4
Left: Schematic illustration of the plasma concentration-time course (solid grey line, C_p), the pharmacological effect (dotted red line) and the biophase concentration-time course (dashed grey line, C_e). **Right**: Schematic illustration of the plasma concentration-time course (solid grey line, C_p), the pharmacological effect (dotted red line) and the biophase concentration-time course (dashed grey line, C_e). The pharmacological response displays saturation (flat upper portion). Note that the biophase concentration-time course peaks at the same time as the pharmacological effect in both plots.

The half-life governing the equilibrium between plasma concentration and response usually falls in the second to minute time range (Table 6.1). The distribution half-lives of cardiovascular effects such as the QT- (duration of the QT wave interval in the human ECG, the prolongation of which may result in serious cardiac arrhythmia) and Monophasic Action Potential Duration (MAPD; guinea-pig model to predict potential human QT liability) responses are typically less than 10 minutes. Different drug target classes may however have very different temporal relationships between their concentration-time and pharmacological response-time courses. Whereas the action of an anti-thrombotic drug effect reflects a rapid response with the half-life to equilibrium $t_{1/2ke0}$ being observed in seconds or minutes, the antipsychotic effect of neuroleptics typically displays a $t_{1/2}$ response of 1–2 weeks. In an effect-compartment (distributional) model the equilibrium occurs relatively rapidly (minutes). In contrast, a delay of 1–2 weeks is hardly explained by an effect-compartment model, but rather due to some other mechanism(s) causing the long time to equilibrium.

Table 6.1
Compilation of some responses and their corresponding plasma concentration-response equilibration half-lives for selected compounds.

The half-life governing the equilibrium is called the equilibration half-life, and usually falls in a second-to-minute time range (Table 6.1). The distributional half-lives of cardiovascular effects such as the QT- and MAPD responses are typically less than 10 minutes.

Response	$t_{1/2ke0}$ (min)	Compound	Reference
FEV-1	7.5	Terbutaline	Ooisterhuis *et al* [1986]
FEV-1	11	Theophylline	Whiting *et al* [1981]
Hypokalemia	11.5	Terbutaline	Jonkers *et al* [1987]
Muscle paralysis	4	d-Tubocurarine	Sheiner *et al* [1979]
Muscle paralysis	4	Vecuronium	Fisher *et al* [1985]
Pain	13	Dihydrocodeine	Webb *et al* [2001]
Spectral edge	1.2	Thiopental	Stanski *et al* [1984]
Spectral edge	6.4	Fentanyl	Scott *et al* [1985]
Spectral edge	1.1	Alfentanil	Scott *et al* [1985]
PR-prolongation	2	Verapamil	Schwartz *et al* [1989]
QT-prolongation	2	Disopyramide	Whiting *et al* [1980]
QT-prolongation	7	Dofetilide	Ollerstam *et al* [2005]
QT-prolongation	8	Quinidine	Holford *et al* [1981]
QT-prolongation	6.4	N-Acetylprocainamide	Piergies *et al* [1987]

Consider the situation where a drug has been administered as a rapid intravenous injection, and the decline of the plasma drug concentration and the rise and fall of the response-time courses have been measured simultaneously (Figure 6.5, left). As can be seen, the plasma drug concentration starts at a high value and then declines throughout the experiment. The response, however, starts at zero, rises to a maximum value and then returns to zero (Figure 6.5, right). There is no apparent relationship between the time course of drug in plasma and the pharmacological response. However, the trick here is to add an effect compartment between plasma and response, which rises from zero, peaks at the same time as response maximum, and then returns to zero. The concentration in the effect compartment is at equilibrium with the pharmacological response at all times. At equilibrium it is assumed the effect compartment concentration to be the same as the plasma concentration, and a first-order rate constant k_{e0}, governs the time to equilibrium between the two compartments. The effect compartment concentration C_e *versus* response relationship gives a picture of the plasma drug concentration-response relationship at equilibrium. The C_e *versus* response is the pharmacodynamic model (E_{max}, EC_{50}, n), and the k_{e0} parameter takes the temporal differences between plasma and response into account. In addition to the pharmacodynamic parameters E_{max}, EC_{50} and n obtained from fitting the model (Equations 6:1–6:2) to response-time data, we also estimate k_{e0}. The underlying assumption is, namely, that there is some kinetic information in the response-time course that allows estimation of k_{e0}. The fitted plasma concentration-time curve combined with the k_{e0} parameter, will give the effect-compartment-time course (second graph from left). Remember that C_e is only a hypothetical curve. There are seldom specific sites in the body where the drug concentration behaves exactly as C_e. Whatever concentration you have of C_e, the pharmacodynamic model will transform that concentration to a pharmacological response (third graph from left) by means of the E_{max}, EC_{50} and n parameters. When the two relationships are combined the predicted response-time course (right hand graph) is obtained. So, by means of the original plasma drug concentration-time course, a pharmacokinetic model (C_p) can be fitted to these and a set of kinetic parameters such as clearance and volume may be estimated.

The concentration profile of the effect compartment is calculated by means of the k_{e0} parameter and Equation 6:1,

$$\frac{dC_e}{dt} = k_{e0} \cdot \left(C_p - C_e\right) \tag{6:1}$$

simultaneously with the pharmacodynamic model (Equation 6:2) that is fitted to the response-time data of Figure 6.5 (right hand panel).

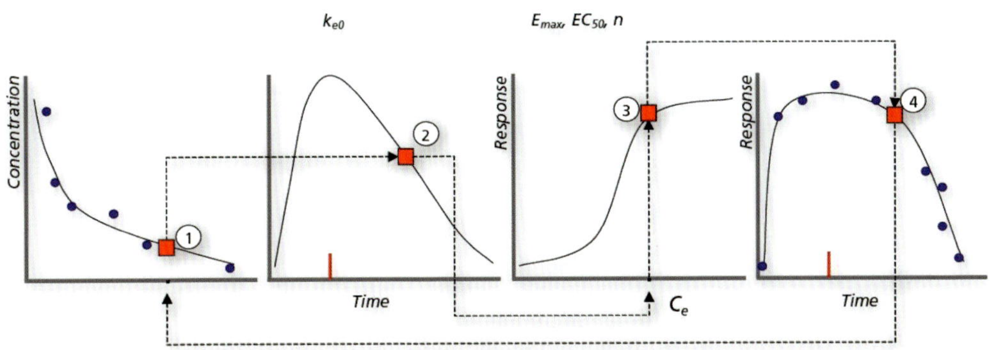

Figure 6.5
Schematic diagram of the effect-compartment model. The drug is given as an intravenous bolus dose. The concentration at time point ① in plasma corresponds to the effect-compartment concentration at ②. The k_{e0} parameter will enable us to generate the C_e *versus* time relationship. The k_{e0} parameter gives us the time delay between C and C_e. The C_e at ② will generate a response corresponding to ③ provided response R *versus* time data are available (④) for estimation of E_{max}, EC_{50} and n. The vertical red bars on the time axes of graphs 2 and 4 shows the same time to maximum effect-compartment concentration and maximum response.

The effect compartment concept (Figure 6.3, A) linked to the Hill equation (E_{max} model, Equation 6:2) may be used for describing such a relationship between drug concentration and intensity of pharmacological effect.

$$Effect = E_0 + \frac{E_{max} \cdot C_e^{\,n}}{EC_{50}^{n} + C_e^{\,n}} \tag{6:2}$$

Here, E_0, E_{max}, EC_{50}, C_e and n denote the baseline effect, efficacy, potency, effect compartment concentration and slope factor, respectively. The E_{max} model is a mathematical representation of the grey *S-shaped* dotted curves in Figure 6.6. The upper left hand plot shows how the width of the hysteresis loop is governed by the effect-compartment rate constant k_{e0}, in that the lower k_{e0} the longer is the time to equilibrium and the wider is the loop. The height of the concentration-response curve is governed by the efficacy of the drug E_{max}. The steepness of the concentration-response curve around the EC_{50} value is governed by n. The larger the n is the steeper is the slope. Recall also from Chapter 3 on plasma protein binding that the steepness and placement of the concentration-response relationship is governed by the free fraction f_u provided total plasma concentrations are used. The positioning of the concentration-response curve along the concentration axis is given by the potency EC_{50}. The higher the potency of a compound, *i.e.*, low EC_{50} value, the more to the left on the concentration axis the concentration-response curve is shifted.

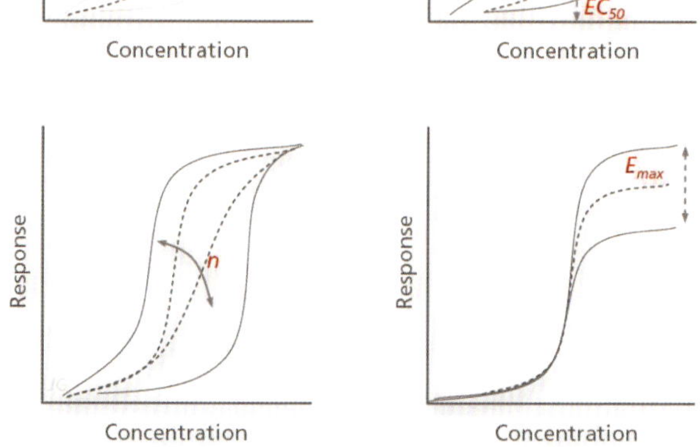

Figure 6.6
Schematic illustration of how the effect-compartment rate constant k_{e0} (upper left) affects the loop-size, potency EC_{50} (upper right) impacts the horizontal placement of the collapsed curve, sigmoidicity parameter n (bottom left) changes the steepness of the concentration-response curve, and the E_{max} parameter (bottom right) impacts the height of the concentration-response relationship.

It is wise to always strive to obtain accurate and precise parameter estimates of k_{e0}, EC_{50}, E_{max} and n. In some cases pharmacological data may not suffice for a full Hill-type relationship. If so, there is a need to reduce the pharmacodynamic model to something simpler such as a linear-, exponential- or log-linear effect model. A lipophilic, as compared to a more hydrophilic or charged, compound may reach the target or biophase more quickly, therefore resulting in a shorter half-life of the distributional process $t_{1/2ke0}$. Processes that occur subsequent to the actual binding to the receptor are related to the physiology/ pathophysiology (system), and are the same for all compounds acting at the same receptor with the same mechanism.

Figure 6.7 shows two common experimental scenarios where complete response data are lacking on the up- and/or down-swing of the response-time course. The first example (Figure 6.7, top figures) is where a single point response-time observation is paired with a single plasma concentration-time point post-response. The other example (Figure 6.7, bottom figure) is where concentration- and response-time data are lacking during the build-up of response.

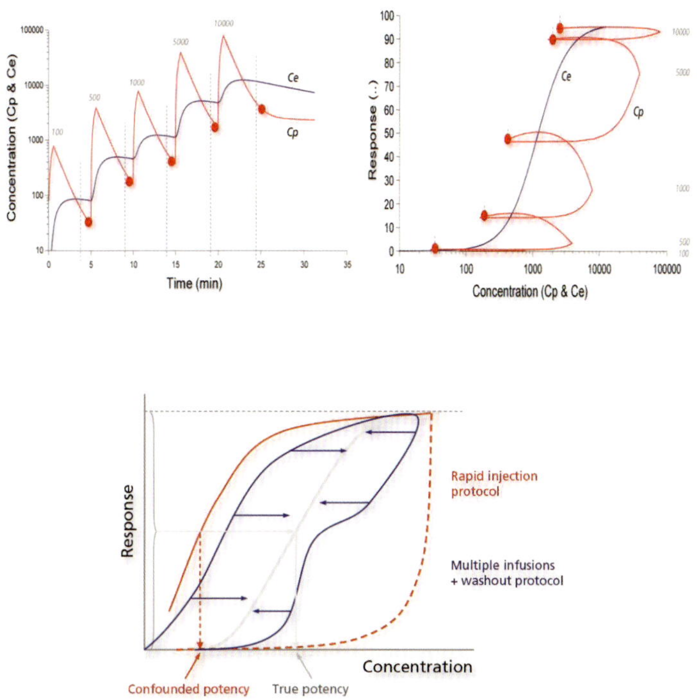

Figure 6.7
Top left: The plasma concentration-time course after repeated rapid injections. The pharmacological measurement is done at the time points indicated by a vertical dashed grey line and the exposure measure at each red dot. Note that exposure is measured after, rather than before, the actual response measurement. The response is a result of the concentration-time course prior to the measurement rather than after. This approach will give an incorrect and too high potency value.
Top right: The simulated concentration-response relationship plotted in time order. Connecting the red dots in this graph will result in a concentration-response curve that is misleadingly shifted to the left *versus* the true concentration-response relationship depicted by the smooth blue line.
Bottom: Schematic illustration of two concentration-response relationships following the post-peak phase of a rapid injection protocol (red solid-dashed curve) and a controlled multiple infusions + washout protocol (blue solid line). The true equilibrium concentration-response relationship is shown by the grey sigmoidal curve in the background. Note also how the estimate of true potency can be confounded by neglecting the time delay between plasma concentration and response.

Although many investigators have used this approach without adopting a rigorous experimental verification procedure, such a 'hit and run' approach was often based on a single dose experiment, with estimates of E_{max} (efficacy) and EC_{50} (potency) being reported in the literature. The effect-compartment model assumes that the lag in pharmacological effect is attributable to distribution (seconds, minutes) of drug from plasma to biophase. However, the model is unsuitable for a fairly large number of different pharmacological effects that are not rate-limited as a result of distribution, but rather because the response occurs over hours, days, or weeks, as exemplified in Sections 6.3 and 6.4. The effect-compartment model also does not allow to focus on specific processes leading to the expression of a pharmacological effect because it cannot discriminate between drug (EC_{50}, E_{max}) and system (physiology, turnover, disease *etc.*) properties.

The data in Figure 6.8 show three scenarios of the application of effect-compartment models. Three different concentration-response relationships are presented as plasma concentration-response loops (red curves), and their corresponding collapsed effect-compartment concentration-response curves (blue). For each case the corresponding terminal log concentration-time and log response-time signatures are also illustrated (inset graphs). In the first case (left) the concentration- and response-time profiles fall in parallel, which means that they are at equilibrium. That tells us that a good estimate of the potency EC_{50} can be obtained from the downswing of the hysteresis curve. During that portion of the curve the system is closest to equilibrium. The yellow line denotes the slope of the concentration-response relationship between the 20 to 80 % response interval.

In the middle graph, the terminal log concentration-time and log response-time do not fall in parallel. The latter response-time profile has a more shallow decline, which suggests a longer response half-life as compared to the plasma concentration half-life. The consequences of this will be a much larger hysteresis loop of the plasma concentration-response relationship. When the hysteresis loop is collapsed, the equilibrium concentration-response relationship will occur closer to the center of the loop. The potency EC_{50} is not obtained from the terminal downswing of the red curve, but rather from the collapsed blue curve in the middle of the loop. Using the red curve for this purpose will give a biased estimate of EC_{50}.

The right-hand graph shows similar data to the first, with only one difference, namely that the slope of the concentration-response relationship is steeper. There is an indication of this in the terminal log-concentration and log-response-time signatures (inset). The response-time course declines more rapidly than the concentration-time course. Remember that the terminal portion of the log-response-time curve may be due to the product of the effect-compartment rate constant k_{e0} and the sigmoidicity parameter n in some instances.

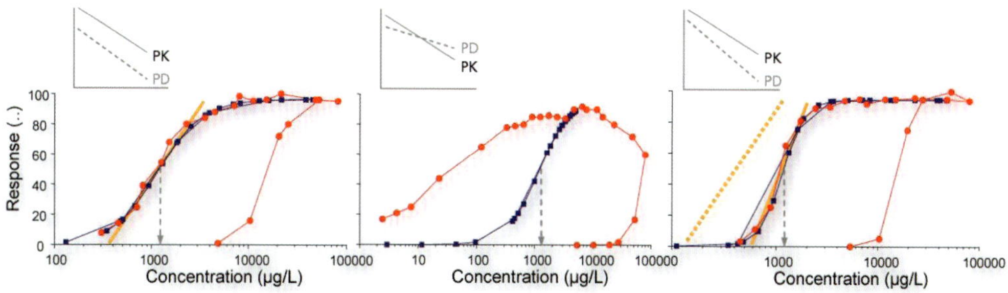

Figure 6.8
Three concentration-response relationships of observed plasma concentrations (red datasets) and predicted effect compartment concentrations (blue datasets) plotted against response. **Left**: Original dataset (Response I). The yellow line denotes the slope of the concentration-response relationship between the 20 to 80 % response interval. **Middle**: A synthetic dataset where all parameters are equal to response I except for the effect compartment rate constant k_{e0} (0.002 min^{-1}). **Right**: A synthetic dataset where all parameters are equal to response I except for the sigmoidicity parameter n (4.1). The dotted yellow line in the right hand plot is inserted for comparison and corresponds to the slope of the log-linear curve in the left-hand plot. The three small insets of concentration and response versus time on a semi-logarithmic scale, Note also the potential for major errors in EC_{50} estimation if only the ascending or the descending portion of the concentration-response curve were to be used.

The terminal portion of the pharmacological response may be approximated to

$$\log(E) \propto .... \ -n \cdot k_{e0} \cdot t \qquad (6:3)$$

Table 6.2
Final parameter estimates and their CV % of responses in Figure 6.8

Parameter	Response I Estimate ± CV %	Response II Estimate ± CV %	Response III Estimate ± CV %
E_{max}	96 ± 1	97 ± 2	95 ± 1
EC_{50}	1160 ± 3	1180 ± 4	1160 ± 2
n	1.9 ± 5	1.6 ± 5	4.1 ± 7
k_{e0}	0.1 ± 6	0.002 ± 2	0.1 ± 5

6.3.2 Effect-compartment analysis of cardiovascular data

In this next example, a new experimental compound is being developed for a cardiovascular disease. From a design point of view five consecutive 30-minute constant rate infusions followed by a washout period is proposed. Data on the plasma concentration, free fraction and cardiovascular effects of compound X from five consecutive intravenous infusions in a pilot study are shown in Figure 6.9 and Table 6.4.

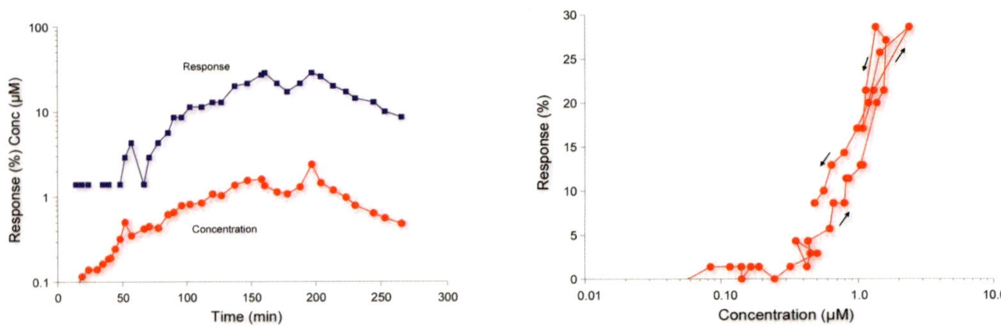

Figure 6.9
Left: Semi-logarithmic plot of observed total concentrations (red symbols) and pharmacological response (blue symbols).
Right: Semi-logarithmic plot of total concentration *versus* response data plotted in time order indicated by the black arrows (right).

Ex vivo plasma protein binding f_u is given in percent (%) and is concentration-dependent (Table 6.4). During the last of the five 30-minute infusions, the syringe jammed for approximately 10 minutes, resulting in a decline in both plasma concentrations and response. The primary model parameters will be the effect-compartment rate constant k_{e0}, that captures the time delay between the plasma concentration-time course and response-time course, and the pharmacodynamic parameters E_{max}, EC_{50}, n and E_0, corresponding to the efficacy, potency, sigmoidicity (exponent) factor and baseline effect, respectively. The plasma concentration necessary to give a 15 percent change in response C_{e15}, will be derived as a secondary parameter. The $T_{1/2ke0}$ and C_{e15} are the primary with respect to design of the next study. $T_{1/2ke0}$ will provide a guide for the establishment of pharmacodynamic steady state, and C_{e15} should be covered on the upswing as well as during the washout.

In this particular exercise a table function is applied for the plasma concentration kinetics. The reason for this is that the infusion was inadvertently interrupted (pump stop) for some time during the 5th infusion, and the resolution of concentration-time data are good. The duration of the stop of the infusion was not recorded and both concentration and response declined during that time period.

$$C_p = f(\,table\,) \qquad (6{:}4)$$

The plasma concentration in terms of a table function *drives* the concentration of drug in the biophase C_e by means of Equation 6:5

$$\frac{dC_e}{dt} = k_{e0} \cdot \left(C_p - C_e\right) \qquad (6{:}5)$$

This equation states that there is a first-order input of drug from the plasma compartment to the effect compartment by means of $k_{e0} \cdot C_p$. There is then a first order loss from the effect compartment by $k_{e0} \cdot C_e$. Because the actual effect compartment concentration C_e was not measured two separate rate constants cannot be estimated for input and output processes. Since the collapsed concentration-response curve displays an *exponential* (or logarithmic) relationship (Figure 6.9, right), Equation 6:6 can be proposed to be an appropriate pharmacodynamic model of an *exponential* concentration-response relationship

$$E = E_0 + a \cdot C_e^b \qquad (6{:}6)$$

where E_0, a and b are the model parameters. Alternatively, a *logarithmic* concentration-effect relationship may be applied

$$E = m \cdot ln(\,A_0 + C_e\,) \qquad (6{:}7)$$

where A_0, and m are the model parameters. Finally, the *Hill* model is fitted to data

$$E = E_0 + \frac{E_{max} \cdot C_e^n}{EC_{50}^n + C_e^n} \qquad (6{:}8)$$

where E_0, E_{max}, EC_{50} and n are the model parameters to be estimated. The *exponential* model (Equation 6:6) has three parameters (E_0, a and b), the *log-linear* model (Equation 6:7) two (C_0 and m), and the *Hill*-model (Equation 6:8) four (E_0, E_{max}, EC_{50} and n) parameters. The concentration-response plot makes it clear that the *Hill*-model is too ambitious because the response does not level off at higher plasma concentrations. Therefore, the E_{max} parameter and indirectly the EC_{50} parameters will be poorly estimated and highly correlated.

Figure 6.10
Left: Observed (black circles) and fitted (solid line) response-time data using a table function of the plasma concentrations that drives the response.
Right: Semi-logarithmic scale of observed (red symbols) unbound concentration-response data. Semi-logarithmic plot of concentration versus response data plotted in time order indicated by the blue arrows. The manually collapsed concentration response relationship is represented by the blue line and the equilibrium concentration-response relationship by the orange line obtained from the regression of the data in the left hand plot.

Concentration- and response-time data are plotted on the left of Figure 6.10 and concentration-response data on the right. Note the slight counter-clockwise hysteresis loop when concentration data are plotted against response data in time order. The time-delay between concentration and response, as seen in the hysteresis loop in Figure 6.10 (right), illustrates the fact that it takes time for the drug to distribute from plasma to the biophase assuming that distribution to and from the biophase is the rate-limiting step. The size of the loop is, however, small so a relativly short equilibration half-life can therefore be expected. By manually collapsing the hysteresis loop in Figure 6.10 (left) a clearer picture emerges of the concentration-response relationship at equilibrium and also an initial estimate of the equilibration rate constant k_{e0}.

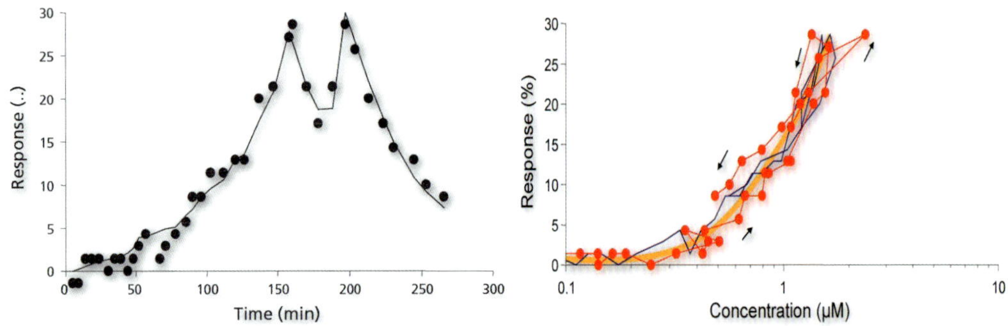

By inspecting the collapsed loop, it is easier to observe what specific concentration results in the response at equilibrium together with the steepness of the relationship and the potential maximum effect. A plateau response, above which there is no further rise in response with increasing plasma concentrations, is not observed in this case. The right hand plot shows the modelled (smooth) relationship between effect compartment concentration and response. The final estimates of the model parameters are shown in Table 6.3 of the *exponential*, *logarithmic* and *Hill*-models.

Table 6.3
Final parameter estimates and their coefficients of variation (%), weighted residual sum of squares *WRSS* and Akaike Information Criterion *AIC*

	Exponential model			Logarithmic model			Hill model			
Parameter	a	b	k_{e0}	m	C_0	k_{e0}	E_{max}	EC_{50}	n	k_{e0}
Estimate	2.1±10	0.6±4	0.078±10	5.2±4	0.6±30	0.079±35	55±30	75±65	0.8±10	0.086±10
WRSS	72			358			61			
AIC	164			224			160			

The *exponential*, *logarithmic* and *Hill* models estimate the plasma concentration corresponding to a 15 % change in response from baseline to 22 ± 5 %, 17 ± 10 % and 24 ± 4 %, respectively and the corresponding distributional half-lives were 8.9, 8.7 and 8.1 min. The *Hill* equation has the lowest weighted residual sum of squares and *Akaike Information Criterion* values, but the poorest parameter precision. In light of the combined goodness of fit criteria and parameter precisions the exponential model is the model of choice. But, remember to be careful with extrapolations beyond the observational range with any of the three models.

Figure 6.11 shows the nonlinear relationship between unbound plasma concentrations *versus* the free fraction f_u measured *ex vivo* in three dogs, and data obtained from an *in vitro* experiment with pooled dog plasma. With the nonlinear plasma protein binding taken into account, it can be observed that there is a shallower unbound concentration-response relationship than that seen in the total concentration-response relationship (Figure 6.12). It should be recalled that total concentrations are ratios of unbound concentrations and free fraction C_u/f_u.

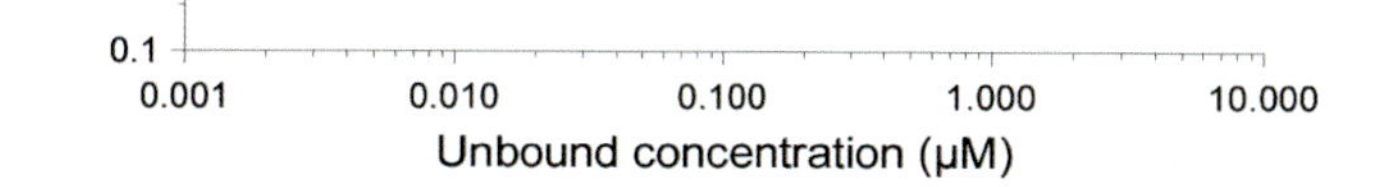

Figure 6.11
Log-logarithmic scale of unbound plasma concentrations *versus* free fraction measured *ex vivo* at 2 to 6 concentrations in three dogs (filled red circles, black diamonds and blue squares). The solid triangle is an *in vitro* measurement of compound X using a commercial kit for assessment of plasma protein binding.

What conclusions can be drawn from the left hand panel of Figure 6.12 when the utility function (difference between positive and negative responses) is incorporated into the analysis? It becomes more obvious that there is no actual benefit from the test compound above the 20 % change in response. The negative effects appears to be substantial already at low unbound plasma concentrations (< 10 nM) where the utility curve passes through zero.

Figure 6.12

Left: Semi-logarithmic plot of total (black symbols) and unbound concentration (constant f_u of 0.46 % blue symbols and *ex vivo* measured f_u shown by red symbols) *versus* response data plotted in time order.

Right: Semi-logarithmic plot of unbound plasma concentrations *versus* observed primary (red symbols), secondary (blue symbols) response data and the utility curve (grey symbols).

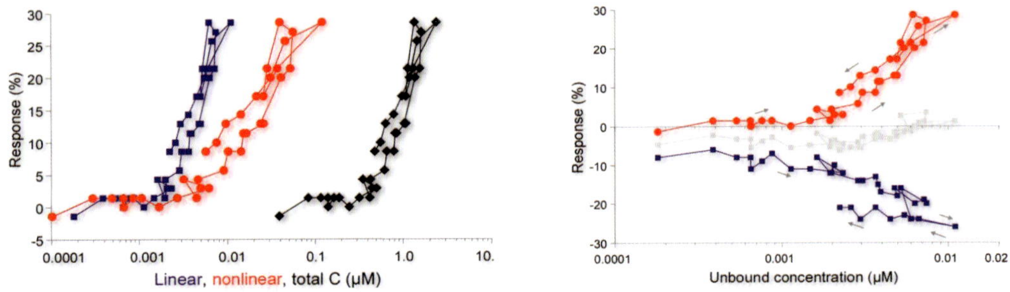

Table 6.4
Time (h), total concentrations (µM), *ex vivo* determined free fraction (%), and QT (% change of baseline) data

A $T_{1/2ke0}$ of seven minutes means that the time to equilibrium between plasma and biophase will take 3–4 times 7 min, which is approximately 20–30 minutes.

Time (h)	Conc (µM)	Free fraction	Effect (..)	Time (h)	Conc (µM)	Free fraction	Effect (..)	Time (h)	Conc (µM)	Free fraction	Effect (..)
5.0	0.00	0.18	-1.4	67.0	0.42	1.05	1.4	160.5	1.35	2.94	28.6
9.0	0.04	0.26	-1.4	71.0	0.45	1.10	2.9	170.0	1.14	2.51	21.4
14.0	0.08	0.35	1.4	78.0	0.43	1.07	4.3	178.0	1.08	2.39	17.1
18.5	0.12	0.42	1.4	85.5	0.62	1.45	5.7	188.0	1.31	2.85	21.4
23.5	0.14	0.47	1.4	90.0	0.66	1.54	8.6	197.0	2.39	5.07	28.6
30.0	0.14	0.47	0.0	96.0	0.79	1.80	8.6	204.0	1.45	3.15	25.7
34.5	0.16	0.52	1.4	102.5	0.82	1.86	11.4	213.5	1.19	2.62	20.0
39.0	0.19	0.57	1.4	111.5	0.85	1.92	11.4	223.5	0.98	2.18	17.1
40.5	0.19	0.58	-	119.9	1.08	2.39	12.9	230.5	0.79	1.80	14.3
44.0	0.25	0.69	0.0	126.5	1.04	2.30	12.9	244.5	0.64	1.49	12.9
48.0	0.32	0.84	1.4	137.0	1.37	2.99	20.0	253.0	0.56	1.33	10.0
52.0	0.51	1.22	2.9	147.0	1.56	3.36	21.4	265.5	0.48	1.17	8.6
57.0	0.35	0.91	4.3	158.0	2.5	3.48	27.1	-			

6.3.3 Effect-compartment analysis of monophasic action potential MAPD

As the focus on "safety margin thinking" increases (for example in handling the hERG/*QT* issue), it is of utmost importance that complex calculations are based on robust experimental design. Essential questions regarding the potency and effective concentration range of a compound are linked to determinations of how much of the compound and what duration of exposure are required for it to produce its desired and undesired effects (onset and intensity) *in vivo*.

Again, this can be exemplified by Case *A* (Figure 6.13). If estimated from direct concentration-effect data in the guinea-pig monophasic action potential duration MAPD QT prediction model, the safety margin is overestimated because of the apparent *hysteresis - i.e.,* there is a lag time between the measured unwanted effect and the measured plasma level (Figure 6.13, Case A). Case B (Fig. 6.13) shows how the safety margin is increased when the temporal difference between plasma concentration and effect for both the primary effect (efficacy) and secondary effect (safety) is taken into account. Using the guinea-pig model, a direct comparison was made between the monophasic action potential duration MAPD potency of the wellknown QT-prolonging agent Moxifloxacin, applying either a repeated rapid injection protocol (non-optimal protocol) or a multiple constant rate intravenous infusion + washout protocol (optimized protocol). The former protocol resulted in the plasma concentration measured at a 10 % C_{e10} (ms) change in MAPD from baseline of 6 μM, whereas the optimized protocol taking *hysteresis* into account resulted in a C_{e10} of 24 μM. This is an important consideration when designing protocols for safety pharmacology.

Figure 6.13
Schematic illustration of safety margin for Case A where the primary effect (efficacy) is compared to the secondary effect (safety) after an infusion+ washout regimen of test compound. Data were only obtained during the infusion. Case B shows a scenario with hysteresis in both the primary and secondary effect data. Safety data based on a rapid injection regimen where data are obtained only from the washout phase. The apparent safety margin seems to be underestimated.

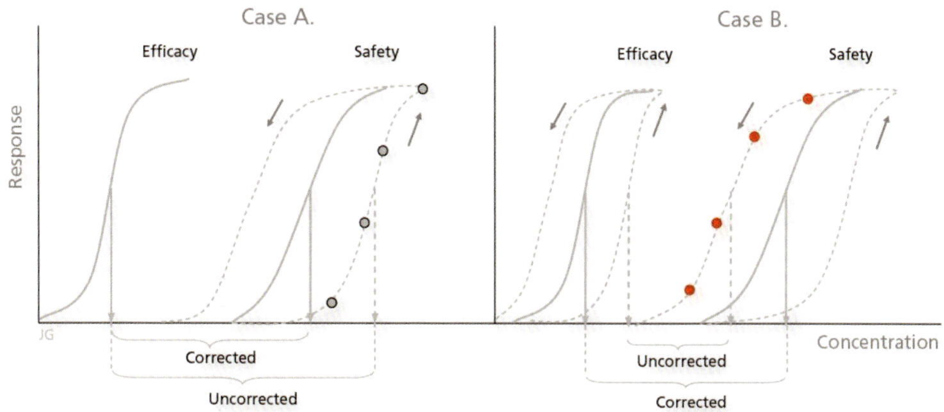

Uncertainty in the target level of efficacy needed for a clinical effect (*e.g.,* 20, 50 or 80 % of maximal effect) and the corresponding plasma concentration range (EC_{20}, EC_{50} or EC_{80}) also affects the assessment of a safety margin. It becomes even more difficult when comparing efficacy and toxicity on an *in vitro/in vivo* basis, particularly when encountering bioanalytical problems for highly bound compounds, large species-dependent differences in binding or concentration-dependent binding. The temporal differences often seen between plasma concentration and pharmacological response have to be tackled for a correct assessment of the primary pharmacological effect and safety margin. Thus, the following Case Study is aimed to show how to analyze real data that by their origin differ in their concentration-time and response-time courses.

A new experimental compound is being developed, and data on the plasma concentration and MAPD effects of compound X in a guinea pig model are collected (Figure 6.14 and Table 6.6) according to the multiple intravenous infusions + washout approach. Plasma protein binding f_u is about 1.5 %.

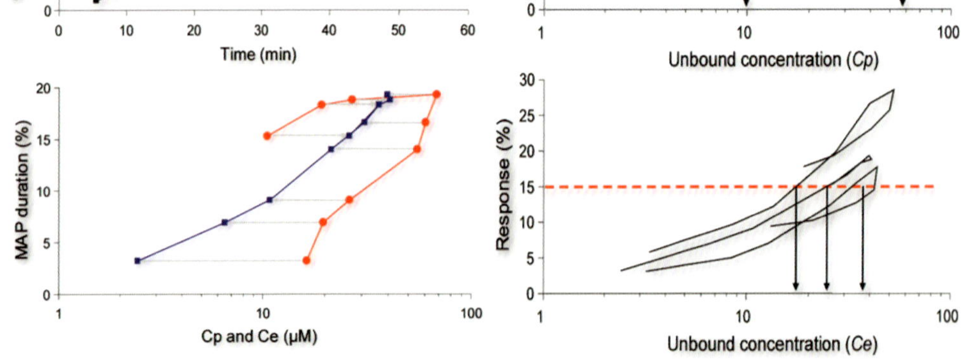

Figure 6.14
Concentration- and response-time course of compound X in animal no. 3 after a dual consecutive intravenous infusion regimen coupled to washout (upper left). Concentration-response curves of all animals after the multiple infusions + washout regimen. (upper right, data plotted in time order). Concentration-response data of animal no. 3 plotted in time order (red curve) and after having collapsed (blue curve) the data (bottom left). Collapsed concentration-response data of all animals (bottom right).

The first infusion regimen was approximately 1 μmol·kg^{-1} for 16 min and the second regimen approximately 2 μmol·kg^{-1} operating between 16 and 32 minutes, which was then followed by a washout period. Now examine the data in Table 6.6 and contemplate what kind of exploratory analyses should be used and what kind of pharmacokinetic and pharmacodynamic model is appropriate? Data of all animals were then initially analyzed by means of a *model-independent* approach based on collapsing plasma response time-course of the compound.

The collapsed concentration-response curves in Figure 6.14 (bottom right) suggest that an effect compartment model (Equation 6:9) may suffice to estimate the temporal differences between plasma concentration and pharmacological response.

$$\frac{dC_e}{dt} = k_{e0} \cdot \left(C_p - C_e\right) \tag{6:9}$$

The effect-compartment model will then be coupled to an exponential pharmacodynamic model (Equation 6:10). By this we mean an *exponential* concentration-response relationship at equilibrium.

$$E = E_0 + a \cdot C_e^b \tag{6:10}$$

where E_0, a, C_e and b are the baseline value, proportionality constant, the effect compartment concentration and exponent, respectively. An *exponential* model is proposed due to the fact that the collapsed data in Figure 6.14 (bottom right) display an erectile shape - this despite a logarithmic X-axis. One of the goals of the safety study is to estimate the plasma concentration of test compound at certain target response levels. The plasma concentration C_{e15} that gives rise to a 15 % change in response is estimated by means of Equation 6:11

$$C_{e15} = \left(\frac{15 - E_0}{a}\right)^{\frac{1}{b}} \tag{6:11}$$

Replacing the target response value of '15' by '10' or '20' yields the corresponding target plasma concentrations.

Another alternative to the exponential model is a *log-concentration-response* model (Equation 6:12)

$$E = m \cdot \log\left(C_e + C_0\right) \tag{6:12}$$

where m, C_e and C_0 are the slope of the *log-concentration-response* curve, the effect compartment concentration and the hypothetical concentration giving rise to a baseline response, respectively. Using the *logarithmic concentration-response* relationship, the expression for plasma concentration C_{e15} that gives rise to a 15 % change in response, is then

$$C_{e15} = e^{\left(\frac{15}{m}\right)} - C_0 \tag{6:13}$$

The reason for the apparent temporal delay, observed as counter-clockwise hysteresis in the concentration-response plot, is most probably a slow distribution of drug from plasma to biophase and back again. The counter-clockwise hysteresis means that the response is lagging after the plasma concentration (See Figures 6.1 and 6.2). The same plasma concentration at an early and late time point results in a low and high response, respectively. When the hysteresis is collapsed it is easier to observe the true concentration-response relationship, because collapsed loop shows the equilibrium relationship. The shape of that relationship reveals the steepness, placement (e.g., potency EC_{50}) and the eventual maximum response. In other words, the relationship between the plasma concentration and response has not been confounded by their different time courses. It is easier to assess potency and efficacy and compare different concentration-response relationships from different compounds by means of the collapsed concentration-response relationship, when the temporal differences are removed.

Hysteresis can be due to distributional delays, or delays caused by a turnover rate-limited response, or because of a slow on/off-rate at the receptor. A distributional delay can be caused by a slow transfer of test compound between plasma and the biophase and back to plasma again. Distributional delays usually fall in the region of seconds or minutes, but are very seldom in the hour range. Counter-clockwise hysteresis means that the plasma concentration peaks earlier than the response. This can be modelled by means of an effect-compartment model. Therefore, the quest of the effect compartment approach is to remove the hysteresis, by introduction of the first-order distributional rate constant k_{e0}, allowing easier observation of the equilibrium relationship between plasma concentration and response. The collapsed data of the three animals displayed an *erectile log-concentration-response* shape, which suggests an *exponential* pharmacodynamic model. No concentration-response plateau was seen.

The final parameter estimates are shown in Table 6.5 together with the secondary parameter C_{e15}. The latter had high precision. The half-life of the equilibrium process ranged between 10 and 15 min. The plasma concentrations giving rise to a 15 % response are 15, 22 and 35 µM.

Table 6.5.
Final parameter estimates and their corresponding CV % for each animal

Parameter	Animal no. 1	Animal no. 2	Animal no. 3
E_0	-3.6	-1.3	-5.0
k_{out} (min^{-1})	0.07	0.07	0.04
a	4.4	1.8	5.2
b	0.51	0.62	0.43
C_{e15}	17	35	22

Does modelling carried out using mean data work? No, totally unrealistic parameter estimates would be obtained, with very low precision, if an analysis on naïve pooled data were to be performed. Pooling and averaging data prior to fitting is to be discouraged due to the pronounced risk of biased results. In the new design longer exposure-response-time courses, including extended washout periods and higher doses, should be undertaken.

Table 6.6
Animal number, time (h), total concentrations (µM) and MAPD (% change from baseline) data

Animal no.	Time (min)	Conc (µM)	Effect (%)	Animal no.	Time (min)	Conc (µM)	Effect (%)	Animal no.	Time (min)	Conc (µM)	Effect (%)
1	0	0		2	0	0		3	0	0	-
	5.5	14.5	5.8		5.5	14	3.1		5.5	16.4	3.2
	10.5	18.8	9.7		10.5	18.8	5		10.5	19.7	6.9
	15.5	23.2	12.2		15.5	20.3	7		15.5	26.3	9.1
	17	-	-		17				17	-	-
	22.5	54.2	19.8		22.5	51.2	12.3		22.5	55.3	14
	27.5	64.8	26.6		27.5	49.6	15.7		27.5	60.5	16.6
	32.5	77.3	28.5		32.5	62.5	17.7		32.5	68.3	19.3
	34	-	-		34	-	-		34	-	-
	39.5	30.7	25.7		39.5	27.5	14.5		39.5	27	18.8
	44.5	21.8	23		44.5	18.6	12.7		44.5	19.3	18.3
	54.5	16.1	19.3		54.5	10	10.2		54.5	10.5	15.3
	64.5	13.8	17.8		64.5	7.9	9.4		64.5	8.7	-

6.4 Delays caused by turnover

6.4.1 When is a turnover model needed?

The lack of mechanistic description of temporal differences between concentration and a pharmacological response by means of the effect-compartment model, is the primary reason why *mechanism-based turnover models* have rejuvenated the field, permitting identification of the physiological components of drug action that are affected by disease, drugs, gender, age, species differences, and other covariates (*cf.* Figure 6.3, B, above). Turnover models also permit separation of drug parameters (EC_{50} and E_{max}) from parameters that govern the physiology or disease (turnover rate and half-life of effect). This means that there are three distinct properties to tackle in a turnover type of model, namely the *physiology* (system, k_{in}, k_{out}), the *pharmacology* (EC_{50} and E_{max}) and the *pharmacokinetics* (*Dose, Cl, V*). Ideally, the latter two are drug specific and can be modified in the drug molecule, whereas the physiology is distinctly different from the drug properties. The physiological properties are something that may be studied by 'challenging' the time course of a biomarker by means of dosing with the drug. Typical physiological properties of a turnover model are the synthesis, secretion or production rate of the biomarker, and the corresponding filtration, catabolism or loss rate of biomarker.

The basic turnover model is based on a one-compartment model with a zero-order production term, denoted the turnover rate k_{in}, and a first-order loss term governed by the fractional turnover rate constant k_{out}. Many physiological and biochemical systems can be approximated by means of the simple turnover model in Figure 6.15.

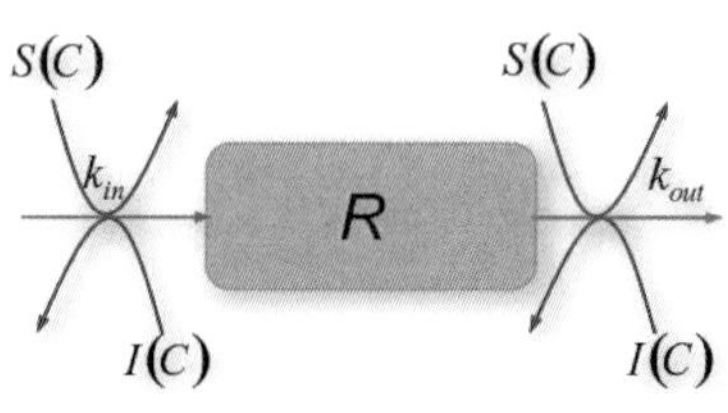

Figure 6.15
Basic structure of a turnover model. The drug may either stimulate (*S(C)*) or inhibit (*I(C)*) the build-up or loss of response. The stimulatory or inhibitory functions are called the 'drug mechanism' functions. A typical inhibitory function is given by Equation 6:16. The turnover rate k_{in} and fractional turnover rate k_{out} are respectively the zero-order and first order rate constants that govern the pharmacological response *R*.

The turnover of response *R* (any kind of biomarker or clinical effect) can then be described mathematically by means of Equation 6:14 where the rate of change of response is governed by production (secretion, synthesis) and loss (catabolism, filtration).

$$\frac{dR}{dt} = production - loss = k_{in} - k_{out} \cdot R \tag{6:14}$$

The production may be synthesis, secretion or any type of accumulation of a biomarker. For example, if the biomarker is some kind of enzyme activity, the more enzyme that is synthesized or produced, the higher is the enzyme level and hence activity. The loss term may be any type of removal of the biomarker, such as filtration, catabolism or transport. The higher the loss rate, the lower the response; provided production or build-up is unchanged. The level of response R is a consequence of flow into and flow out of the 'tank'. When a drug with a stimulatory or inhibitory mechanism acting on either production or loss of response is added, we will eventually see a change in the value of R.

The drug is assumed to act upon either the production k_{in} or loss k_{out} of response, either via an inhibitory or stimulatory drug 'mechanism function'. The latter represents the pharmacological action of the drug. The third part of the model is given by the plasma kinetics of drug which 'drives' the drug mechanism function. A few literature examples of turnover models are shown in Figure 6.16, and a more in-depth discussion is given in Section 6.3.2.

Figure 6.16
Basic structure of a turnover model. The drug may either stimulate or inhibit the build-up or loss of response. The turnover rate k_{in} and fractional turnover rate k_{out} are respectively the zero-order and first order rate constants that govern the pharmacological response R.
Case A: Warfarin inhibits the regeneration of Vitamin K and thereby block the production of the Prothrombin Complex Activity R.
Case B: Furosemide stimulates the diuresis via blocking the reuptake of water in the tubuli.
Case C: Erythropoietin Epo stimulates the production rate of the red blood cells.
Case D: CB1-antagonists stimulates loss of body fat (and thereby body weight), as a consequence of their appetite-reducing and increased energy expenditure action.

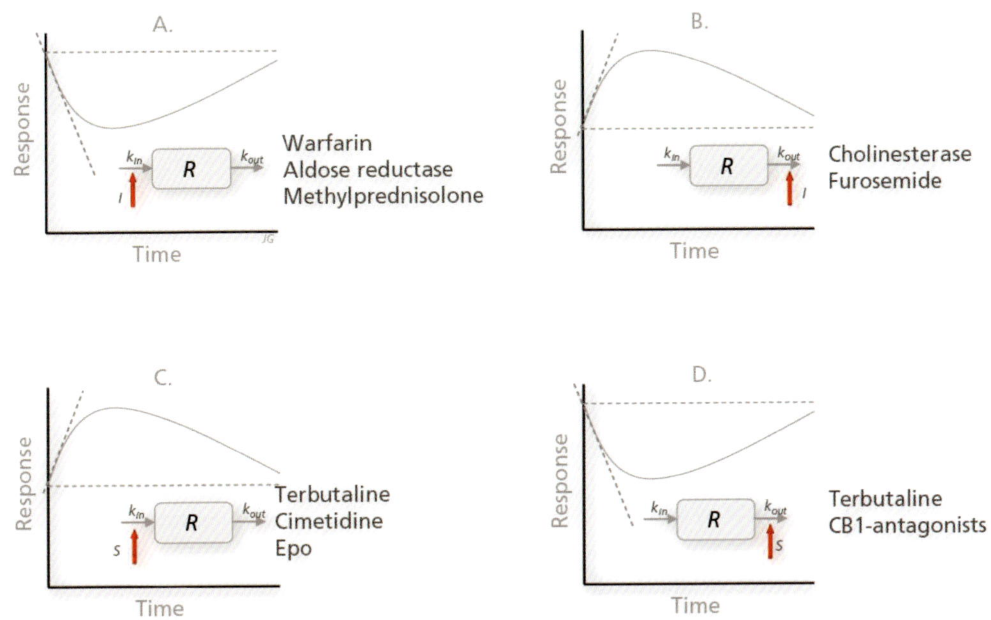

6.4.2 Inhibition of enzyme activity

It should be remembered that there are three distinct properties to tackle when setting up a turnover type of model, namely the *physiology* (system, k_{in}, k_{out}), the *pharmacology* (EC_{50}/IC_{50} and E_{max}/I_{max}) and the *pharmacokinetics* (*Dose, Cl, V*). In this Section we show the:

- Response data on GSECRE activity from mice brain
- Exploratory analysis of experimental data
- Output from fitting a mechanistic turnover model to data
- Interpretation of pharmacodynamic parameters
- Superiority of time-series analysis.

This can be illustrated by a typical turnover-driven response with compound X, a gamma secretase, *GSECRE*, inhibitor. *APP* mice produce excess human *Amyloid Precursor Protein APP*, which following cleavage initiates a cascade of secretases including *GSECRE* and generates fragments that eventually deposit to form insoluble plaques. By inhibiting *GSECRE* the objective is to lessen plaque formation in the brain via a decrease in the formation of amyloid peptide fragments. The concentration of soluble *beta-amyloid* $[A\beta_{1-40}]$ (which is called 'response' from now on) in brain serves as a biomarker of enzyme activity, which is illustrated schematically in Figure 6.17. An increase in the plasma test compound concentration leads to a decrease in enzyme activity and lower amounts of soluble beta-amyloid are formed. This process is relatively rapid. However, the formation of insoluble *beta-amyloid plaques* occurs over several weeks or months, which is now the rate limiting step for stopping disease progression.

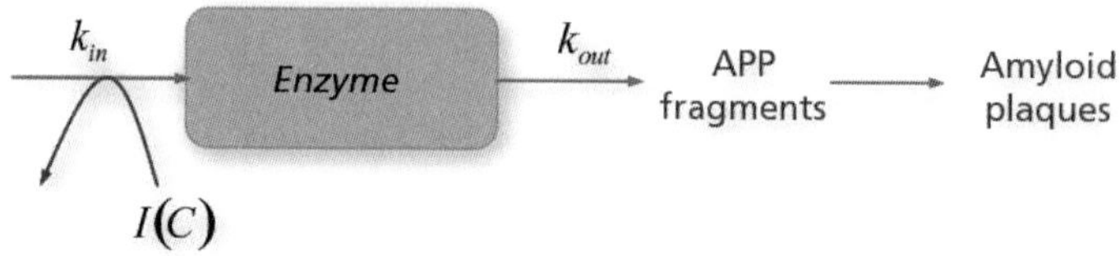

Figure 6.17
Basic structure of a turnover model. The test compound inhibits the enzyme activity or the build-up of response. The turnover rate k_{in} and fractional turnover rate k_{out} are respectively the zero-order and first order rate constants that govern the enzyme activity. The formation of amyloid plaques is a slow process that occurs over weeks or months.

There is a temporal delay between the maximal plasma concentration of test compound X and the maximum decrease in response (Figure 6.18) so there is a situation similar to that shown in Figure 6.1 (right). The peak plasma concentration occurs before the time of maximal intensity of response.

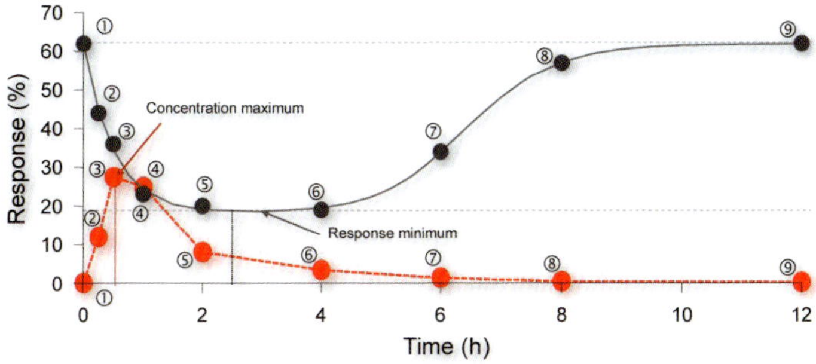

Figure 6.18
Response-time course of enzyme activity and concentration-time data of test compound X. Note that plasma concentration peaks at 30 min, but the time of maximal effect (response trough) occurs at about 2.5 h. The information about the maximal intensity of response at the plasma peak concentration is not optimal.

The concentration-response relationship shows a clear time-delay in terms of hysteresis, a flat minimum response region in spite of a 8-fold concentration range (3 to 25 μM), and then a steep return towards the baseline response (Figure 6.19). The encircled numbers in Figures 6.18 and 6.19 correspond to the concentration and response-time data combined in chronological time order. The response starts at the baseline ① and gradually declines as the plasma concentration increases. The increase in plasma concentration more effectively inhibits the enzymatic action of *GSECRE* thereby lowering the production and consequently the response level. This graph displays a pronounced hysteresis between the onset and offset of response when plotted against plasma concentration. At time point ③ maximum plasma concentrations have been reached, but the maximum intensity of response has not. In spite of a drop in plasma concentrations from 25 to 5 μM between time points ④ and ⑥, very little change in the intensity of response is seen. This portion of the loop shows saturation, which suggests an inhibitory I_{max}- type of drug mechanism function (Equation 6:15).

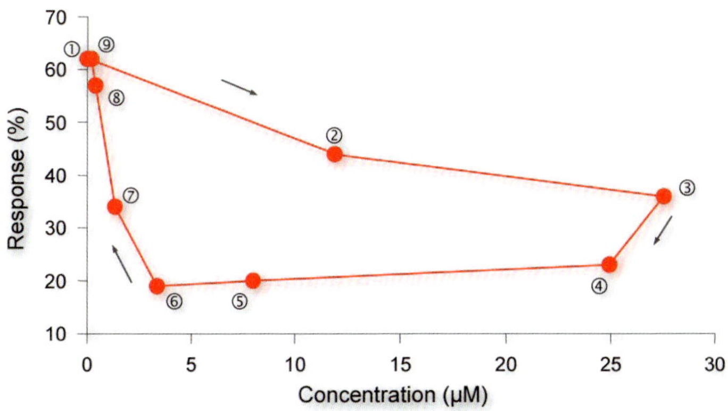

Figure 6.19
Concentration-response relationship of compound X in mice. Note the clockwise hysteresis or time-delay between plasma concentrations of compound X and the pharmacological response. The numbers represent experimental data plotted in time order starting with the baseline ①, reaching concentration maximum ③ and response maximum (curve minimum) ⑥ before the response returns back to the baseline ⑨. An early exposure of, for example 10 μM (upper portion of the curve) exhibits very little reduction of response, whereas the same concentration (10 μM) at some later time point produces a near maximum inhibitory response (bottom portion of the curve).

A first-order one-compartment model was fitted to the plasma concentration-time data. This curve (Equation 6:15) only serves as a smooth fit of the concentration-time data to *'drive'* the inhibitory 'drug mechanism' function (Equation 6:16).

$$C = A \cdot \left[e^{-K \cdot t} - e^{-K_a \cdot t}\right] = 32.1 \cdot \left[e^{-0.566 \cdot t} - e^{-14.2 \cdot t}\right] \qquad (6:15)$$

The kinetic parameters (A, K and K_a) were then fixed to their estimated values (32.1, 0.566 and 14.2) and that function then served to drive the *'drug mechanism function'* ($I(C)$, Equation 6:16) when fitting the turnover model to response-time data.

$$I(C) = 1 - \frac{I_{max} \cdot C^n}{IC_{50} + C^n} \qquad (6:16)$$

$I(C)$ means that the inhibitory action is a function of the plasma drug concentration C. I_{max} and IC_{50} denote the efficacy and potency to be estimated.
The turnover of response can be described mathematically as

$$\frac{dR}{dt} = production \cdot inhibition - loss = k_{in} \cdot I(C) - k_{out} \cdot R \qquad (6:17)$$

The basic turnover model contains two parameters, the turnover rate k_{in} and the fractional turnover rate k_{out}. The turnover rate, also called production in Equation 6:17, is inhibited by the *'drug mechanism function'* $I(C)$ (Equation 6:16). Initially, when there are high concentrations of test compound and we therefore observe maximum inhibition ($I(C) \approx 0$)of the production term, Equation 6:17 can be approximated by

$$\frac{dR}{dt} \approx -loss = -k_{out} \cdot R \qquad (6:18)$$

Because the inhibition of production is initially large, resulting in a low product $k_{in} \cdot I(C)$ term, the loss term $-k_{out} \cdot R$ will dominate the turnover process, which then results in a first-order rate of decline of response. This loss is similar to first-order elimination $-K \cdot C$ of drug in a one-compartment model.

The slope of the downswing of the response-time course on a semi-logarithmic scale, corresponds to $-k_{out}$, which is estimated as

$$slope = -k_{out} = \frac{\ln(R_1) - \ln(R_2)}{t_1 - t_2} = \frac{\ln(60) - \ln(35)}{0 - 0.5} \approx -1\,h^{-1} \tag{6:19}$$

At minimum response, the rate of change is zero, which is given by Equation 6:20

$$\frac{dR}{dt} = k_{in} \cdot I(C) - k_{out} \cdot R_{min} = 0 \tag{6:20}$$

If Equation 6:20 is rearranged and solved for R_{min}, then

$$R_{min} = \frac{k_{in}}{k_{out}} \cdot I(C) = \frac{k_{in}}{k_{out}} \cdot (1 - I_{max}) \tag{6:21}$$

The distance Δ between an approximate baseline R_0 of 60 units and minimum response R_{min} of 20 units is 40 response units, which corresponds to the product of $R_0 \cdot I_{max}$. The distance Δ between basline R_0 and R_{min} is written mathematically as

$$\Delta = R_0 - R_{min} = \frac{k_{in}}{k_{out}} - \frac{k_{in}}{k_{out}} \cdot (1 - I_{max}) = \frac{k_{in}}{k_{out}} \cdot I_{max} \tag{6:22}$$

The baseline R_0, minimum response and Δ are known, which means that also I_{max} can be estimated to 0.3. Recall that the baseline R_0 is written as

$$R_0 = \frac{k_{in}}{k_{out}} \approx 60\ \% \tag{6:23}$$

which after rearrangement gives a k_{in} equal to

$$k_{in} = R_0 \cdot k_{out} \approx 60 \cdot 1 = 60\ \% \cdot h^{-1} \tag{6:24}$$

The IC_{50} parameter of the 'drug mechanism function'

$$I(C) = 1 - \frac{I_{max} \cdot C^n}{IC_{50} + C^n} \tag{6:25}$$

is obtained from 50 % of response on the upswing of the response-time curve (Figure 6.20). The reason for choosing the upswing rather than the downswing of response-time data is that it is assumed to be closer to terminal equilibrium. The plasma concentration at that time point gives an approximate value of IC_{50} (~1μM < 1.33 μM). The exponent of the plasma concentration in Equation 6:25, usually falls in the 1–4 range.

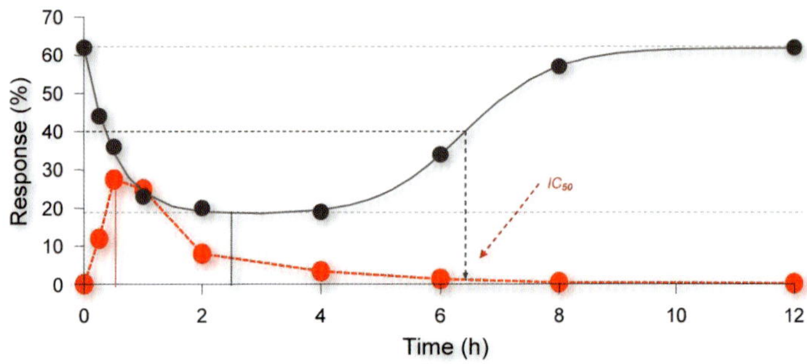

Figure 6.20
Schematic illustration of how potency IC_{50} is estimated from half-maximal distance of maximum response on the terminal response-time course. The concentration indicated by the dashed grey vertical line, is a reasonable estimate of IC_{50}. since plasma concentration and response have reached a pseudo-equilibrium state.

By fitting the combined *'drug mechanism function'* (Equation 6:16) and *turnover* model (Equation 6:17) to the response-time data in Figure 6.18 it is possible to estimate potency IC_{50}, fractional turnover rate k_{out}, baseline R_0, efficacy I_{max} and steepness of the concentration-response relationship n, respectively. The potency is about 1 μM, the half-life of response ($ln2/k_{out}$) is about 20 min and the steepness is a relatively high 2.5. It is also possible to try to set either the n parameter or the I_{max} parameter to unity and see whether the model fits the data well and parameter precision is high. The real value of the analysis after having carried out a regression of the response-time data, is to simulate the response as a function of the steady-state plasma concentration. This can be done by deriving the steady-state expression of response R as a function of the steady-state plasma concentration C.

$$R_{ss} = \frac{k_{in}}{k_{out}} \cdot I(C) = R_0 \cdot \left(1 - \frac{I_{max} \cdot C_{ss}^n}{IC_{50}^n + C_{ss}^n}\right) \tag{6:26}$$

Note that response starts at the baseline value R_0 and then gradually decreases down toward a lower minimum value R_{min} as concentration of inhibitor C increases. The relationship between response and the plasma concentration is shown in Figure 6.21 together with the final parameter estimates and their precision. Remember, the higher the plasma concentration the higher is the inhibitory effect on enzyme activity, and consequently the lower is the production of *APP* fragments. This leads to less amyloid plaques which is a much slower process and occurs over weeks and months.

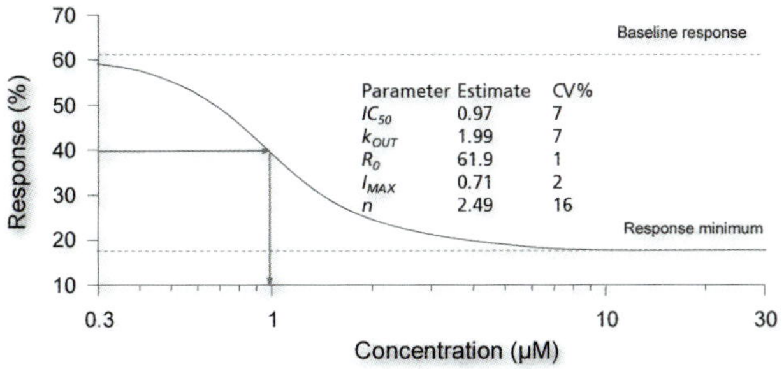

Figure 6.21
Semi-logarithmic plot of the concentration-response relationship of compound X. The final parameter estimates of the system and drug parameters together with their individual precisions (CV %) are also included. Note that this equilibrium function (Equation 6:26) lacks the hysteresis loop when plotted. That is because Equation 6:26 mimics the steady-state situation where time is no longer an issue.

Simulations of the response-time course following a low and a high dose are superimposed over the original data from the above analysis in Figure 6.22. Note that the onset of action is not affected by a change in dose as the time of observation was at C_{max}. There is very little change in intensity with the other simulated doses. However, the duration of response is increased proportional to the log(*Dose*).

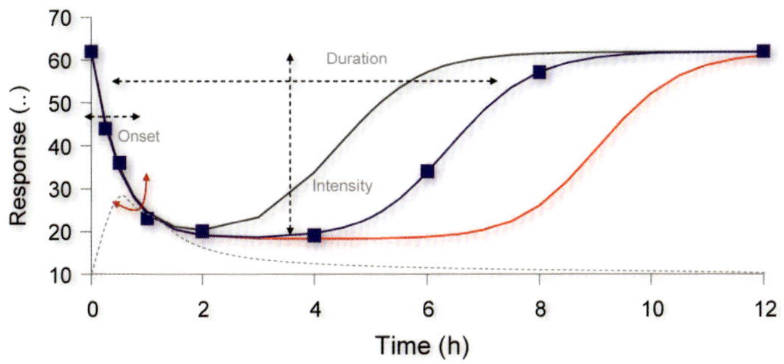

Figure 6.22
Observed and simulated response-time data. Note that the onset of response is the same for all three doses whereas the duration changes. Blue curve: original response data from Figure 6.18; black and red curves depict low and high-dose simulations, respectively, of response-time courses.

If an attempt is made to fit a *link* model (*distributional* model, see Section 6.2) to the data, systematic deviations can be seen between observed and predicted responses (Figure 6.23). This demonstrates that in spite of a single response-time profile (one dose), the link model is inferior to the *turnover* model. The shapes of their respective response-time courses are different because the models are not structurally equivalent.

Figure 6.23
Observed and model predicted response-time data. The blue curve is the original *turnover model* and the red curve the distributional (link) model.

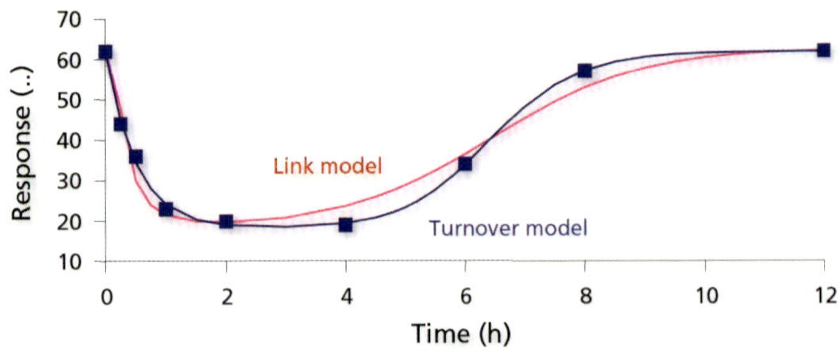

6.4.3 Summary of turnover models

The construction of a *mechanistic turnover model* can be summarized by stating that the *pharmacokinetics* PK and *pharmacology* PD are separated from the *physiology* or system properties (Figure 6.24). The concentration-time course is a function of the dose, the final pharmacokinetic parameter estimates and time. The predicted time course then drives the *'drug mechanism function'*, which is inhibitory in this case. In the Case Study above an enzyme inhibitor was examined. This particular inhibitory *'drug mechanism function'* has three parameters, namely IC_{50}, I_{max} and n. The functional form is dependent on the mechanism of action and the quality and resolution of response-time data. The third part of the model is how the drug 'mechanism' function influences the actual turnover of response. The turnover is often governed by two parameters k_{in} and k_{out}. So there are five model parameters that are estimated from fitting the three-piece model to the response-time data. Note that turnover is still an ongoing process also when no drug is present.

Figure 6.24
Schematic illustration of the three building blocks pharmacokinetics, pharmacology and physiology of the basic *turnover model*. The pharmacokinetic function includes time, dose and the pharmacokinetic parameters such as clearance *Cl* and volume of distribution V_{ss}. The pharmacological function, also called the drug mechanism function, includes the potency EC_{50}/IC_{50} and efficacy E_{max}/I_{max} parameters. The physiology or pathophysiology includes turnover rate k_{in} and fractional turnover rate k_{out}. Practical real life Case Studies of turnover modeling are collected and analyzed in Gabrielsson and Weiner (2006).

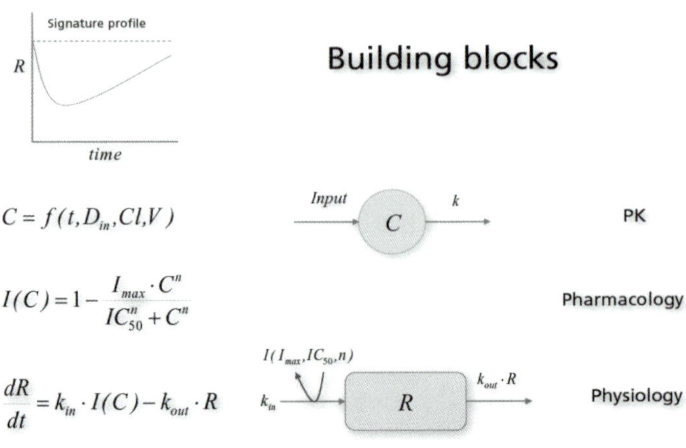

$$C = f(t, D_{in}, Cl, V)$$

$$I(C) = 1 - \frac{I_{max} \cdot C^n}{IC_{50}^n + C^n}$$

$$\frac{dR}{dt} = k_{in} \cdot I(C) - k_{out} \cdot R$$

6.5 Delays caused by slow receptor on/offset

6.5.1 When are receptor on/off models needed?

The third approach to analyzing temporal differences between plasma concentrations and pharmacological response is by means of *receptor on/off binding models.* These models are valid when the ligand-receptor association and dissociation processes are the rate-limiting steps along the dose-concentration-response axis. Many physiological actions are mediated by a transient formation of binary ligand-receptor complexes. Signal transduction depends on agonist engagement by cell-surface G-protein-coupled receptors, while gene transcription depends on binding of transcriptional activators to specific promoter regions of the gene. For the majority of these biological activities, the binary complex between receptor and ligand has a limited lifetime, and the functional consequences of this complex are similarly transient and can be tightly regulated with respect to onset, intensity and duration of response. A schematic illustration of the drug-receptor interaction is given in Figure 6.25 together with the equilibrium relationship between drug ligand concentration and binding or response.

Inactive form *Active form*

$$[Drug] + [Receptor] \underset{k_{off}}{\overset{k_{on}}{\rightleftharpoons}} [Drug\text{-}Receptor] \longrightarrow E$$

$$Response \propto Binding = \frac{B_{max} \cdot [C]}{K_d + [C]}$$

Figure 6.25
Schematic illustration of the drug-receptor equilibrium process. After the drug-receptor complex [Drug-Receptor] has been formed, a pharmacological effect *E* has the potential to develop. The relationship between the drug ligand concentration [C] and binding occupancy or response is given as the mathematical function below. The dissociation constants K_d, is the ratio of the dissociation rate constant k_{off}-to-associaction rate constant k_{on}.

In many physiological situations, the duration of response caused by the receptor-ligand interaction is directly related to the residence time of the binary complex *[Drug-Receptor]* itself. Therefore, the longer the time a ligand binds to the receptor, the longer is the duration of response. Table 6.3 gives some examples of receptors, ligands and residence times.

Receptor	Ligand	Residence time	Indication	Reference
Ribosome complex	erythromycin	1.6 min	bacterial infection	Lovmar *et al.*, 2004
Thrombin	hirudin	14 h	thrombosis	Stone & Hofsteenge, 1986
Dopamine D_2	haloperidol quetiapine	58 min 0.4 min	schizophrenia	Kapur & Seeman, 2000
Epidermal growth factor	lapatinib	7.2 h	HER2+ breast cancer	Wood *et al.*, 2004
Steroid 5a- reductase	finasteride	> 43 days	benign prostatic hyperplasia, prostate cancer and androgenetic alopecia	Bull *et al.*, 1996

Table 6.3.
Receptors, ligands and recidence times of different pharmacological systems (Adapted from Tummino and Copeland [2008])

Since the dissociation constant K_d is actually the ratio of the dissociation rate constant k_{off} and the association rate constant k_{on}, it contains no information about the time course of the build-up and degradation of the drug-receptor complex. In order to demonstrate this situation, simulations are shown for two compounds with equal K_d values but different dissociation rate constants k_{off}. The slower onset and offset of binding occupancy (solid line) is due to the smaller k_{off} value, although both compounds exhibit the same affinity for the receptor at equilibrium (Figure 6.26).

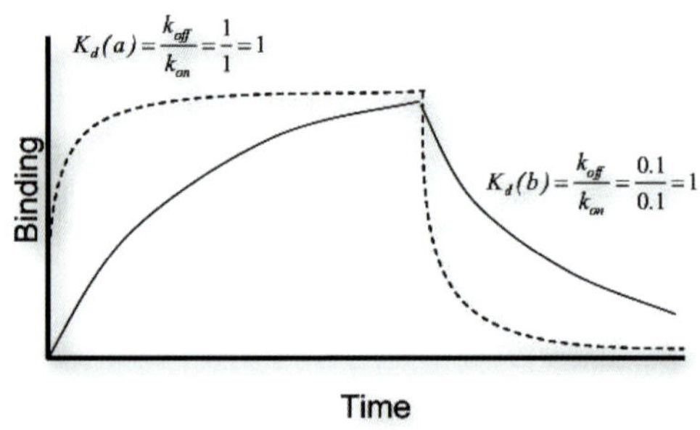

Figure 6.26
Schematic simulation of two response time courses with the same dissociation constants K_d, but with a 10-fold difference in the individual first-order off-rate constant k_{off}. Note the slower onset and offset (b, solid line) using the smaller k_{off} (0.1 time^{-1}) rate constant. Both compounds display the same affinity for the receptor at equilibrium (same K_d), but their binding kinetic parameters (k_{on}, k_{off}) differ by a factor of 10.

To fully appreciate the *in vitro* estimate of potency and the impact of *Target Residence Time* for the assessment of potency and effect duration via its drug dissociation constant K_d, information is also needed about the first-order rate constant k_{off}. Figure 6.27 shows a simulation of the pharmacological response of two compounds that have the same dissociation constants (K_d *i.e.*, ratios of k_{off}/k_{on} are the same) but a 10-fold difference in the k_{off} values, and the consequences on effect duration. It is assumed that the target residence time ($1/k_{off}$) is the rate-limiting step for build-up and loss of response and that all other distributional and physiological processes occur more rapidly.

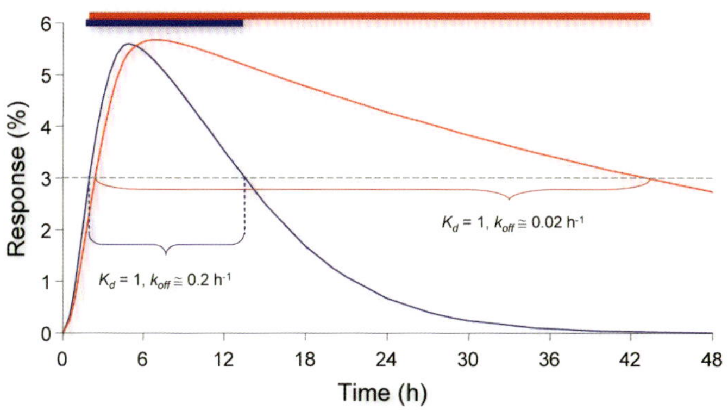

Figure 6.27
Schematic simulation of two response time courses with the same dissociation constants K_d, but with a 10-fold difference in the individual first-order off-rate constant k_{off}. Note the longer duration (red bar) for the time course using the smaller k_{off} (0.02 min^{-1}) rate constant.

The *in vitro* measurement of the dissociative half-life $t_{1/2koff}$ is a crucial metric of compound optimization which may differentiate compounds, leads or series, and can be a key indicator of *in vivo* effect duration and efficacy. Rational, deliberate optimization of residence time may provide novel opportunities to achieve better drug principles, encompassing both extended duration of action and improved selectivity by mitigating against secondary and safety/toxicity events mediated by off-target interactions at supra-therapeutic exposures. For a further detailed, insightful discussion, please see Copeland [2011].

The key driver of the *in vivo* receptor-ligand lifetime is the dissociation rate constant k_{off}. The reciprocal of k_{off} corresponds to the biological complex residence time or duration.

6.5.2 Case Study of delay caused by receptor on/offset

This Section highlights the behaviour of receptor occupancy data, the importance of dose selection and the use and abuse of pharmacodynamic models. When analyzing these data it should be borne in mind that the next step will be to design a multiple oral dose study.

Two dogs received intravenous infusion regimens of 10 (dog 1) and 100 and 800 (dog 2) μmol·kg^{-1} of compound X and receptor occupancy data were obtained over time. In the first part of the analysis, plasma samples were analyzed by means of a one-compartment first-order disposition model. Occupancy-time data are shown in Figure 6.28.

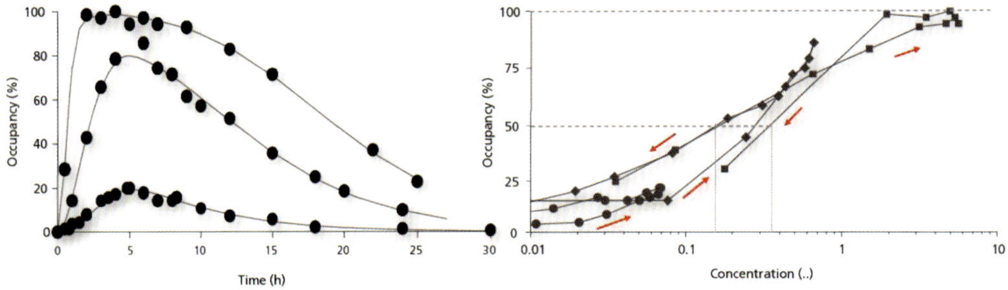

Figure 6.28
Left: Observed (symbols) and model predicted (solid lines) occupancy data of compound X in two dogs after a constant intravenous infusion regimen. The low dose data (10 µmol·kg-1) originates from one dog and the intermediate (100 µmol·kg-1) and high dose (800 µmol·kg-1) from a second dog.
Right: Observed binding occupancy data plotted versus predicted equilibrium plasma test compound concentrations using an effect-compartment approach. The lack of successful collapse is due to fact that an effect-compartment approach based on first-order on/off response does not mimick a second-order on, coupled to a first-order off, process.

The pharmacokinetics of compound X was described by Equation 6:27

$$\frac{dC}{dt} = input - elimination = \frac{Infusion}{V} - \frac{Cl}{V} \cdot C = \frac{Infusion}{V} - k \cdot C \qquad (6:27)$$

The volumes and rate constants are different for the two dogs. The *occupancy-time* data were modeled by means of a *receptor on/off binding* model as described by Equation 6:28

$$\begin{aligned} \frac{d[RC]}{dt} &= on\ rate - off\ rate = \\ &= k_{on} \cdot C \cdot (B_{max} - [RC]) - k_{off} \cdot [RC] \end{aligned} \qquad (6:28)$$

R, C, k_{on}, k_{off} and B_{max} represent, respectively, the response, plasma concentration, receptor-ligand *on rate,* receptor-ligand *off rate* and maximum (proportional to *R*) occupancy.

This shows that there is a nonlinear relationship between drug concentration and the pharmacological response and that a nonlinear response-time model is needed. This is taken care of by the second-order build-up $k_{on} \cdot C \cdot (B_{max} - [RC])$ term coupled to a first-order degradation term in Equation 6:28 when fitting the receptor *on/off* model. The overall fit of the three datasets was good (Figure 6.28) and with high parameter precision. The behavior of Equation 6:28 is shown by the black solid lines in Figure 6.28 for each dose.

The steady state solution to Equation 6:28 is

$$[RL]_{ss} = \frac{k_{on} \cdot C \cdot B_{max}}{k_{off} + k_{on} \cdot C} = \frac{B_{max} \cdot C}{\frac{k_{off}}{k_{on}} + \frac{k_{on} \cdot C}{k_{on}}} = \frac{B_{max} \cdot C}{K_d + C} \qquad (6:29)$$

where K_d is the binding dissociation (affinity) constant and B_{max} the total receptor pool concentration. The full occupancy-time series analysis by means of Equation 6:28 gave a K_d value of 0.19 µM.

Model	k_{off}	k_{on}	B_{max}	K_d
Equation 6:28	0.19	0.96	101	0.19

6.6 Comparisons of the link-, turnover- and receptor on/off models

6.6.1 Background to the Case Study

This Case Study demonstrates the application of the *link-*, *turnover-* and *receptor on/off* models in a real life situation of temporal delays. It also challenges the traditional single time-point assessment of a pharmacological response with the time-series approach. This means that several measures of the pharmacological response are obtained over time, ideally at two or more doses of the test compound. Target engagement is the summary of the dose size, pharmacokinetics and exposure to drug, affinity and efficacy and the physiological reaction (such as amount and turnover rate of the target protein) to the drug-receptor interaction.

In an earlier phase of this project, the experimentalist had measured the pharmacological response at only one time point, namely at 60 min. At that time, this approach was considered to be the most practical since a large number of compounds could be screened in a very short period of time. The 60 min design was aimed at addressing whether there was target engagement or not. Recall that *target engagement* is a consequence of the dose, plasma drug exposure time course (exposure), the affinity and efficacy of the ligand (pharmacodynamics) and the turnover behaviour of the physiological response *per se* (biomarker response). However, we need to dissect the pharmacokinetics, pharmacodynamics and biomarker response from each other in order to fully characterize the onset, intensity and duration of a pharmacological response. Consequently not everybody was convinced about the single 60-min sampling *hit-and-run* approach, and it was proposed that an alternative technique be used which would include more extensive response-time courses at each dose level. In Step 1 a dataset containing two concentration-response-time courses was analyzed by means of an effect-compartment (link) model, a turnover model and a receptor on/off binding model. In Step 2 this dataset was subsequently extended with a third response-time series, to demonstrate which category of models best explained the data.

Data were obtained from a study in dogs after two oral doses of 1 and 2 μmol·kg^{-1} of antagonist X (Figure 6.29 and 6.30). No fitting of the plasma concentration-time data was performed, because it was assumed that substantial and systematic deviations between observed and model-fitted plasma concentrations (Figure 6.30) would not improve the fit of response-time data.

Figure 6.29
Left: Concentration-time data after oral dosing of 1 and 2 μmol·kg^{-1} to a dog.
Right: Response-time data from the same animal.

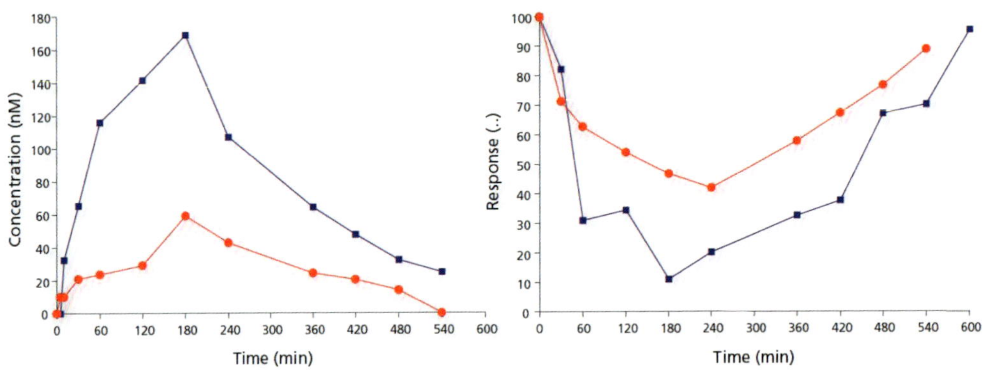

When an oral dose of antagonist X was ingested, it was absorbed from the intestine into the blood stream over a time course typically reaching maximal exposure C_{max} between one and several hours. The absorption process displayed high intra-individual variability in both the rate and extent of absorption. Since no good model representation was found for the plasma exposure, the observed plasma concentration-time data was used by means of a *Table function*. Figure 6.30 (right) illustrates why a *Table function* of the observed plasma concentrations was used rather than fitting a smooth curve to the data. There are systematic deviations between observed and model predicted concentrations. Initially the smooth curve misses the absorption pattern, and terminally it predicts higher concentrations than observed. The latter results in a prolonged estimated duration of response.

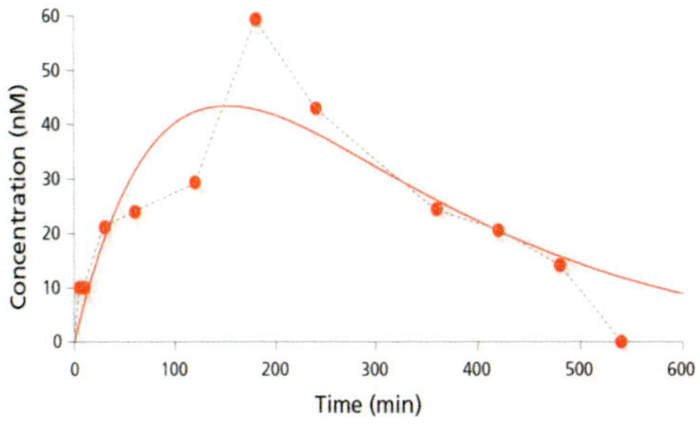

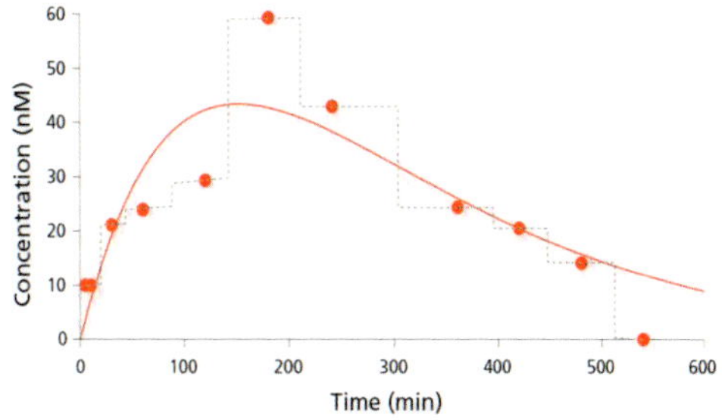

Figure 6.30

Left hand plot: Concentration-time data after oral dosing of 1 μmol·kg^{-1} to a dog. The filled circles are experimental data connected with a grey dashed line, and solid red line is model predicted data of a one-compartment first-order input and output model.

Right hand plot: This figure illustrates the behavior of the table function (grey dashed line) that is applied rather than fitting a function to the experimental concentration-time data. Since there is no smooth representation of the jumpy experimental data, the approach of the table function is preferred.

Figure 6.31 illustrates the three classes of models that were fitted to the two response-time courses in Step 1. In Step 2 only the turnover model will be applied.

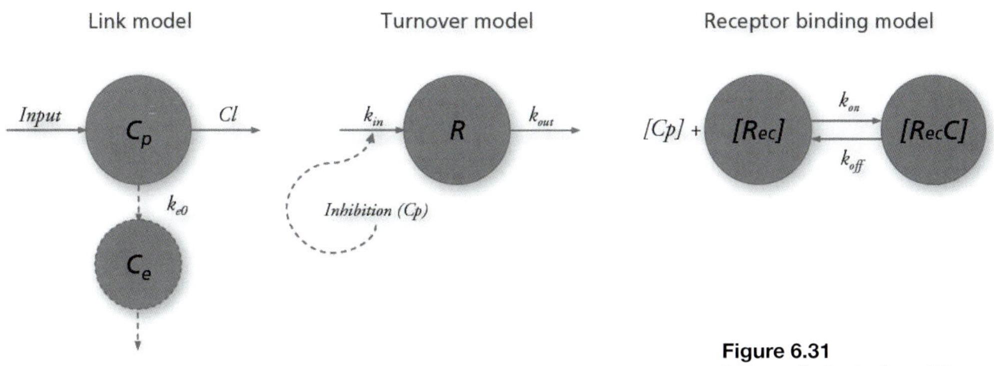

Figure 6.31
Schematic illustration of the *link model* (left), *turnover model* (middle) and *receptor binding model* (right) that will be fitted to the experimental data. Plasma concentration will be implemented as a table function for all three models.

Initial parameter estimates are derived manually using both graphical techniques and the semi-compartmental approach. When the first step of the analysis uses two datasets, the same analysis should be carried out on three datasets simultaneously.

6.6.2 Distributional (link or effect-compartment) model

In this particular exercise a table function for the plasma concentration kinetics was applied because trying to fit an ordinary first-order input/output model gave rise to substantial and systematic devations between the observed and predicted data. A table function uses the observed concentration-time data rather than the predicted pharmacokinetic function such as the first-order input/output function.

$$C_p = f(\,table\,) \tag{6:30}$$

Equation 6:30 drives the disposition of drug in the biophase C_e which is given by *distributional model*

$$\frac{dC_e}{dt} = k_{e0} \cdot \left(C_p - C_e \right) \tag{6:31}$$

Since the collapsed concentration-response curve displays an I_{max}-type (or logarithmic) relationship (Figure 6.32), Equation 6:32 is an appropriate start as a concentration-effect relationship

$$E = E_0 \cdot \left(1 - \frac{C_e^n}{IC_{50}^n + C_e^n} \right) \tag{6:32}$$

where k_{e0}, E_0, IC_{50} and n are the first-order rate constant for the drug disposition of the effect compartment, baseline, potency and sigmoidicity parameter, respectively. Equation 6:32 assumes a total inhibition $I_{max}=1$ of drug when plasma concentration approaches infinity.

6.6.3 Turnover model

The same *Table function* (Equation 6:30) as in Model 1 is also used to drive the *turnover model*

$$\frac{dR}{dt} = k_{in} \cdot I(\,C\,) - k_{out} \cdot R = k_{out} \cdot R_0 \cdot I(\,C\,) - k_{out} \cdot R \tag{6:33}$$

where k_{in}, $I(C)$, k_{out} and R_0 are the turnover rate, drug mechanism function, fractional turnover rate and baseline response, respectively. We re-parameterize the turnover rate by means of $k_{out} \cdot R_0$, because this has been proven to be a more robust parameterization than using k_{in} for the turnover rate.

The inhibitory *'drug mechanism function'* is written as

$$I(C) = 1 - \frac{C^n}{IC_{50}^n + C^n} \quad (6:34)$$

which relates how effectively the plasma exposure of the antagonist is able to inhibit the turnover rate of response. The plasma concentration of antagonist is still given by Equation 6:30 as a *Table function*. Note that when plasma concentration C becomes large, the inhibitory function $I(C)$ approaches zero, which in practice means total blockage and response R is eventually reduced to zero.

The action is *'direct'* or instantaneous on the production, which means that it is *'indirect'* on the response *per se*. The response is indirectly controlled by production and loss (Figure 6.15 and Equation 6:14).

6.6.4 Receptor binding model

The same *Table function* (Equation 6:30) of C as in the distributional (link) and turnover models, drives the *receptor binding model*

$$\frac{d[RC]}{dt} = k_{on} \cdot [C] \cdot [R] - k_{off} \cdot [RC] \quad (6:35)$$

where k_{on}, C, R, $[RC]$ and k_{off} are the second order association-rate constant, plasma concentration of test compound, the free receptor pool, receptor-ligand complex and first-order dissociation rate constant, respectively. The concentration of free, unoccupied, receptors R is expressed as the total receptor pool B_{max} minus the concentration of bound receptors $[RC]$. This expression is written as maximum receptor concentration $B_{max} - [RC]$

$$\frac{d[RC]}{dt} = k_{on} \cdot [C] \cdot (B_{\max} - [RC]) - k_{off} \cdot [RC] \quad (6:36)$$

The resulting model that is fitted to the effect-time data is written as

$$E = 100 - [RC] \quad (6:37)$$

6.6.5 Some results of the analysis

Concentration-response data are shown in Figure 6.32 for the two doses. Figure 6.32 (right) also contains collapsed concentration-response data. By collapsing the hysteresis loop one obtains a picture of what the concentration-response relationship looks like at equilibrium. The plasma concentration that results in a 50 % reduction in response can be read off from the graph. This is an IC_{50} value of approximately 40 nM. In parallel to an approximate estimate of IC_{50} , an approximate number of the first-order equilibration rate constant k_{e0} is also derived from collapsing the hysteresis curve. In this case a value of 0.02 min^{-1} is used. This is also a good starting point for the fractional turnover rate constant k_{out} and the dissociation rate constant k_{off}. The k_{e0}, k_{out} and k_{off} are all first-order rate constants, with their corresponding half-lives obtained as ln(2) divided by the rate constant.

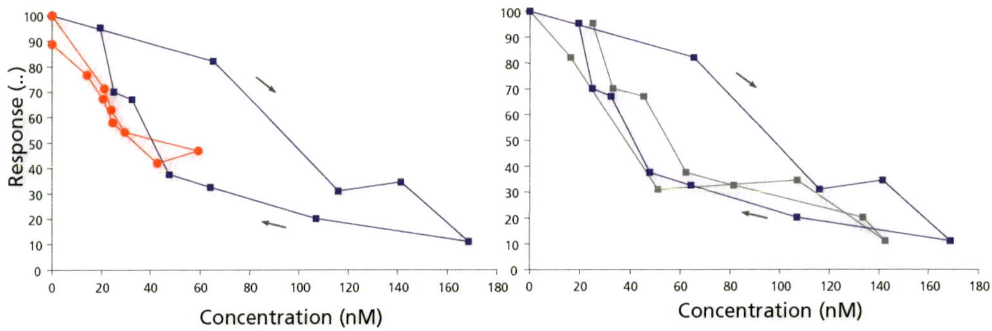

Figure 6.32
Left: Concentration-response data after oral dosing of 1 and 2 µmol·kg^{-1} to a dog. **Right**: Original (blue curve) and collapsed (grey curve) response-time data from the same animal after the 2 µmol·kg^{-1} dose.

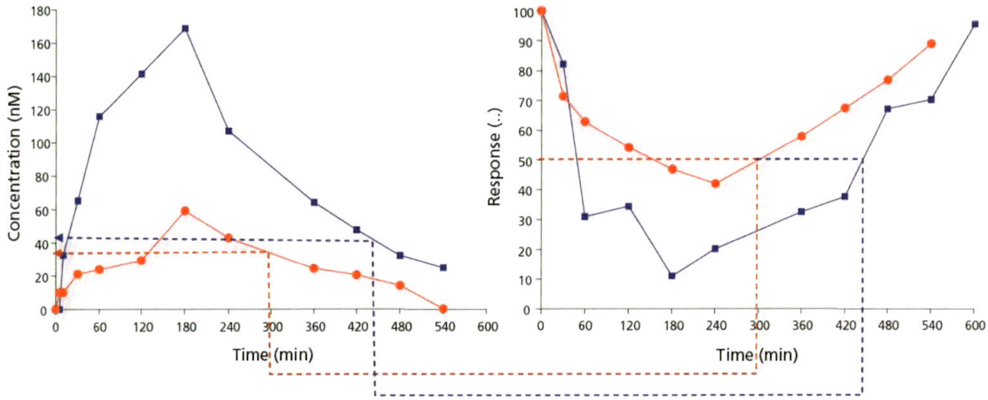

Figure 6.33
Left: Concentration-time data after oral dosing of 1 and 2 µmol·kg^{-1} to a dog. **Right**: Response-time data from the same animal. Start at 50 % of maximum response and then read off the corresponding time on the response-time graph. This time point is then used to get the actual plasma concentration that corresponds to the initial estimate of IC_{50} for the two doses.

An alternative approach to obtain an initial estimate of IC_{50} is shown in Figure 6.33. From the 50 % reduction of response, the corresponding time may be read off during the terminal portion of the response-time course. Those time points correspond to a plasma concentration (left hand graph) which in turn gives an approximate IC_{50} value of 35 to 45 μM. This approach is valid if the assumption is made that the response can be totally blocked by a high dose of the test compound.

A third approach of initial estimates of IC_{50} is by the use of Equation 6:38. The equation is set to zero at the highest dose nadir value (approximately 10) and the corresponding plasma concentration (approximately 160 nM) inserted at that time point, and k_{out} of 0.02 min^{-1} used.

$$\begin{aligned}\frac{dR}{dt} &= k_{out} \cdot R_0 \cdot \left(1 - \frac{C^n}{IC_{50}^n + C^n}\right) - k_{out} \cdot R \\ &= 0.02 \cdot 100 \cdot \left(1 - \frac{160^n}{IC_{50}^n + 160^n}\right) - 0.02 \cdot 10 = 0\end{aligned} \tag{6:38}$$

Assume a value of the exponent n to be 1 and rearrange the equation and solve for IC_{50}. The IC_{50} parameter then becomes approximately 20 nM which is lower than the predicted value, but which can still be used as a reasonable initial estimate. The second-order rate constant k_{on} is estimated from the slope of the downswing of the response-time course of the highest dose

$$-\,slope = \frac{\Delta R}{\Delta t} = -k_{on} \cdot \bar{C} \cdot R_{50} = -k_{on} \cdot 150 \cdot 35 = -0.7 \tag{6:39}$$

where $\bar{C}$ and R_{50} are the average concentration and response within that time interval. The k_{on} parameter then becomes 0.000133 min^{-1}·nM^{-1}.

The experimental and model-predicted response-time data showed high consistency (Figure 6.34). The final parameter estimates and their precision are shown in Table 6.4. The first-order rate constants of the *link-* k_{e0}, *turnover-* k_{out} and *receptor* k_{off} models are 0.025, 0.031 and 0.018 min^{-1}, respectively. The corresponding half-lives then become 28, 22 and 39 minutes. The respective potency values for the three models are 42, 40 and 48 nM.

Figure 6.34
Response-time data after oral dosing of 1 and 2 µmol·kg^{-1} to a dog. Experimental data (solid symbols) and *turnover model* predicted (solid lines) response-time courses. The lack of smoothness in the predicted response-time curves is due to the application of a Table function for the plasma concentration-time courses (Figure 6.30, right).

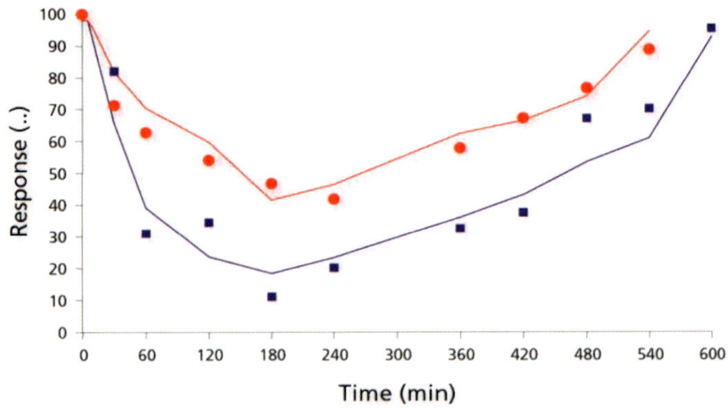

Table 6.4
Final parameter estimates and their CV % from fitting the *effect-compartment model*, *turnover model* and *receptor k_{on}/k_{off} model* to 1 and 2 µmol·kg^{-1} response-time data. K_d is estimated as a secondary parameter and not part of the actual regression

The difference in the half-lives of the turnover and receptor binding models is almost 100 %. The expected duration of response will consequently differ substantially between the two models. The predicted time to pharmacodynamic steady-state will be twice as long based on the receptor binding model as for the turnover model, provided plasma concentrations stay below IC_{50} or K_d.

Effect compartment			*Turnover*			*Receptor k_{on}/k_{off}*		
Param	Estimate	CV %	R_0	98	4	k_{on}	0.00037	31
k_{e0}	0.025	24	k_{out}	0.031	22	k_{off}	0.018	26
n	1.1	14	n	1.1	16	B_{max}	107	12
IC_{50}	42	9	IC_{50}	40	13	K_d	48	27

6.6.6 Refining the turnover analysis with an extended dataset

Nevertheless, the traditional 60 min *hit-and-run* practice was still favoured by a few scientists in the project group, who wanted to challenge with an even larger dose and compare their way of reasoning with the modelling approach. A new experiment was therefore designed with administration of an oral dose of 4 µmol·kg^{-1}. Concentration-time data (Figure 6.35) were taken from the same dog examined previously. Response-time data were collected at the same time points as in previous experiments, and plotted *versus* time (Figure 6.36, left) and concentration (Figure 6.36, right).

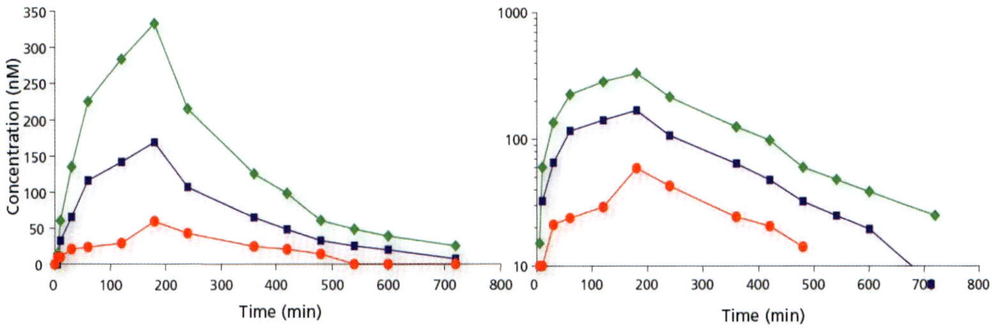

Figure 6.35
Left: Linear plot of concentration-time data after oral dosing of 1, 2 and 4 μmol·kg^{-1} to a dog. **Right**: Semi-logarithmic plot of concentration-time data. Observe how the terminal data fall in parallel.

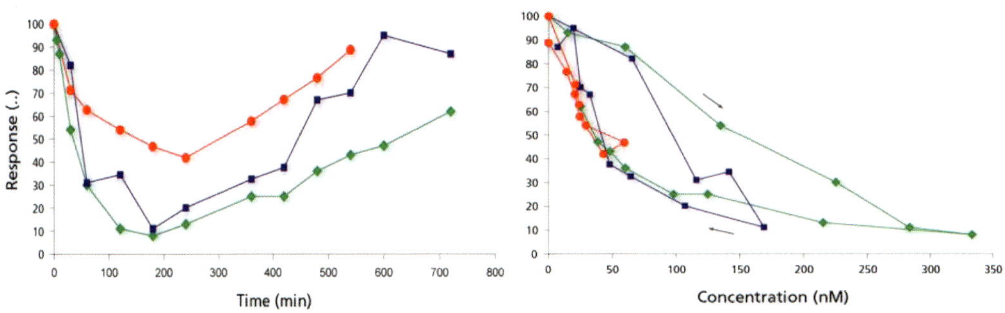

Figure 6.36
Left: Response-time data after oral dosing of 1, 2 and 4 μmol·kg^{-1} to a dog. **Right**: Concentration-response data plotted in time order denoted by the arrows. Observe how the terminal data (ascending phase of hysteresis loops) of the three experiments superimpose where equilibrium is established between concentration and response.

Special attention was paid to the variability seen in data of the two previous experiments. Therefore, a stricter assessment of the biomarker at each time point was introduced. Note that the nadir of response occurs at the same time point as the peak concentration, which suggests a more rapid turnover of response than the plasma kinetics. Previously unused terminal data at 600 and 720 minutes from the 1 and 2 μmol·kg^{-1} were also added (Figure 6.36). Assessment of the hysteresis curve from the highest dose experiment by means of collapsing the loop gave an approximate value of k_{out} of 0.02 min^{-1} (Figure 6.37). This value may also be used as a good initial estimate of the other first order rate constants k_{e0} and k_{off}. Note the clear saturation in response beyond concentrations of 200 nM.

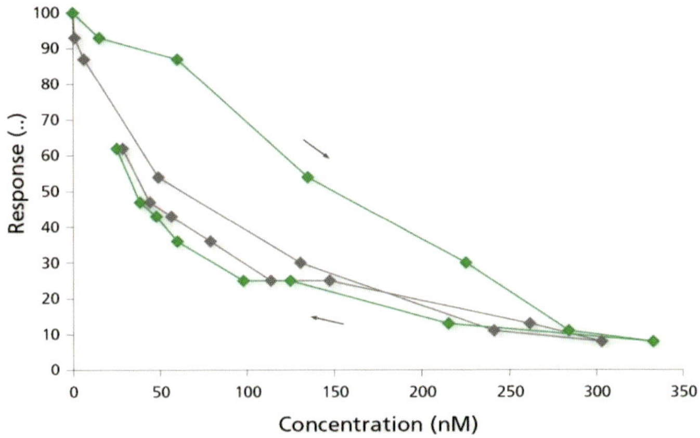

Figure 6.37
Concentration-response data of the highest dose 4 μmol·kg^{-1} to a dog plotted in time order denoted by the arrows (green curve). Collapsed data using an *effect-compartment* rate constant of 0.02 min^{-1} (grey curve).

The same initial estimates of the turnover model were still used when fitting all three response-time courses simultaneously. As you can see in Table 6.5, an increased precision of the turnover model parameters was obtained, although their actual estimates remain the same. Figure 6.38 shows the high consistency between experimental and model predicted values. Since actual plasma concentrations, rather than model-predicted, were used it was difficult to make any judgement about the peak-shift in response with an increase in dose. Figure 6.38 shows that the variability in experimental data from the 4 μmol·kg^{-1} dose experiment was less than for the two lower doses.

Table 6.5
Initial and final parameter estimates and their CV % obtained from fitting the *turnover model* to the extended 1, 2 and 4 μmol·kg^{-1} response-time data

Parameter	Initial estimates	Final estimates	CV %
R_0	100	98	3
k_{out}	.02	0.032	15
n	1	1.2	11
IC_{50}	40	43	9

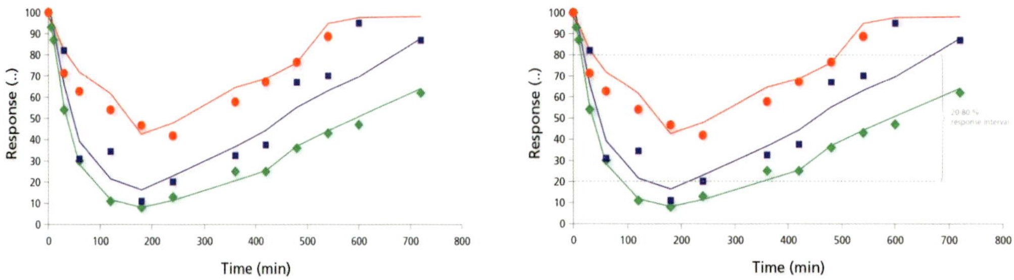

Figure 6.38
Response-time data after oral dosing of 1, 2 and 4 μmol·kg^{-1} to a dog at three different occasions. Experimental data (solid symbols) and model predicted (solid lines). Right hand figure also shows the 20–80 % of maximum response interval where the recession lines run in parallel with a slope equal to $k \cdot m$.

The kinetic-dynamic analysis was then compared to the outcome of the traditional *hit-and-run* practice where concentration-response data after three different doses at 60 min have been superimposed on the complete data set used for the regression analysis (Figure 6.39).

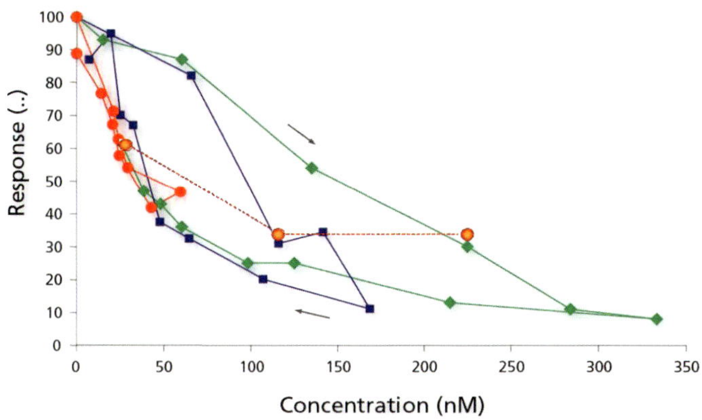

Figure 6.39
Concentration-response data after oral dosing of 1, 2 and 4 μmol·kg^{-1} to a dog at three different occasions. Experimental data (solid symbols) and model predicted (solid lines). Data collated from the traditional 60 min hit-and-run practice are shown as orange symbols connected by a red dashed line.

The *hit-and-run* practice misses maximum drug induced response (intensity of response) by at least 20 response units, and underpredicts the effective concentration range. The onset and duration of response could not be appreciated from a single time point nor could a valid assessment of potency be done. The estimated potencies of the hit-and-run practice and modeling approach are 20 and 43 nM, respectively. This is a 100 % difference caused by bias in the former estimate.

6.6.7 Conclusions about target engagement

In summary, an integrated approach using quantitative methods better categorizes candidate drug profiles, and thus improves the drug screen. The time-series approach discriminates between pharmacokinetic, pharmacodynamic and system properties. A key achievement in this project was the thorough *re-design* of the *in vivo* pharmacological model, which meant abandoning the 60 min *hit-and-run* practice in favour for the time-series approach, resulting in optimized data quality. By combining the time-series data with a turnover model it was possible to make more informed predictions about onset, intensity and duration of response. The collection of both rising and declining response-time and concentration-time data, is from an experimental design point of view the ideal approach to emulate. Target engagement was demonstrated at the lower (<50 nM) nanomolar range, with a half-life of response of 20 min (ln2/0.032).

Chapter 7

Principles of Inter-Species Scaling

7.1 When and why do we extrapolate data across species?

In the previous chapters, principles of pharmacokinetics and pharmacodynamics have been discussed and illustrated by real data and Case Studies obtained from experiments in various species. General pharmacokinetic properties were presented from a pharmacodynamic point of view. Plasma protein binding was specifically and separately covered in Chapter 3. Chapter 4 dealt with nonlinearities and why exposure, and preferably unbound concentrations, are important from a pharmacodynamic perspective. The handling of rapid and time-delayed concentration-response relationships were discussed in Chapters 5 and 6, respectively. The present chapter will show

- How to predict the pharmacokinetic or pharmacodynamic properties, such as dose, clearance and potency in humans, using data obtained in other species
- The impact of variability and uncertainty in parameters such as clearance and potency, and how they are affected by changes in factors such as plasma protein binding
- The combined uncertainty in these parameters and how they alter the human dose prediction as illustrated in a dose nomogram.

The pharmacokinetics of many drugs can be predicted in man from available animal data. However, in some instances animal drug data do not allow accurate predictions to be made. What is interesting to know is why some compounds fail, what makes them different and how can the predictability be improved?

Inter-species scaling is carried out during all drug discovery phases. Early extrapolations are done from an *in silico* model based on physicochemical data in lead identification (LI) or early lead optimization (LO) phase. It is common to extrapolate from *in vitro* metabolic data obtained in microsomes and hepatocytes already during the early lead generation (LG), but also onwards throughout all drug development phases until human data become available. The *in vivo* pharmacodynamic estimates of potency (EC_{50}) are commonly used as predictors of effective concentrations before more robust *in vivo* biomarkers or disease markers are available. Scaling is also done on *in vivo* data obtained from acute intravenous or extravascular dosing in prenomination phase, and on toxicokinetic data obtained during completion of the documentation submitted to support an investigational new drug (IND) or clinical trial application (CTA) to the authorities (Figure 7.1).

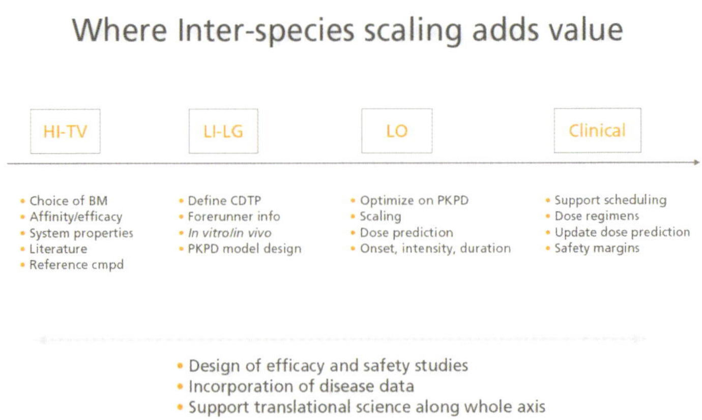

Figure 7.1
Overview of value added inter-species scaling during drug discovery and development. The choice of a relevant biomarker (BM), information of affinity/efficacy, system properties (turnover of biomarker, impact of disease) and information regarding reference compounds guide the process of Hit Identification to Target Validation (HI-TV). Defining the Candidate Drug Target Profile, using forerunner information (if available) and applying *in vitro/in vivo* relationships improves the characteristics of the Lead Compounds identified during Lead Identification (LI) and Lead Generation (LG), and the PK and/or PD parameters to further optimize during Lead Optimization (LO) phase. Kinetic-dynamic reasoning also guides the scaling of PK and PD properties including dose predictions to man during prenomination prior to initial human trials. During the clinical phases, a quantitative pharmacological approach supports study design, dose scheduling, update dose predictions and define safety margins. In parallel to PKPD, inter-species scaling provides the integration instrument for the estimation of the first human dose, the predicted concentration-time course and the response-time course in man.

Obviously, the earlier in drug development that extrapolations are made to man, the more the inherent risk with the predicted value (Figure 7.2). Consequently, extrapolations made later are generally more predictive and confirmative. However, pooling different sources of *in vitro* and *in vivo* data generally increases the understanding of variability in the human predicted parameter. Since intrinsic clearance is obtained from *in vitro* experiments in an artificial physiological context, it is a crude measure of the true hepatic capacity and does not necessarily take non-linearities (such as the V_{max} and K_m parameters) or uptake transport processes into account. Non-linearities, however, are fortunately often disclosed by *in vivo* experiments.

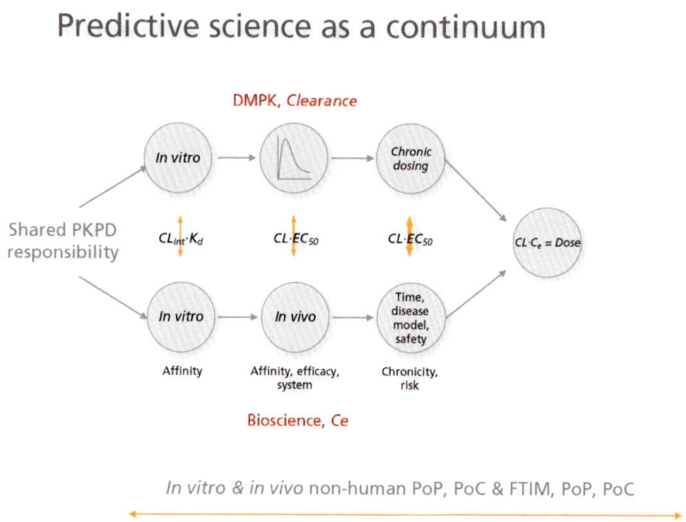

Figure 7.2
Predictive science suggest that ranking of compounds are done based on the combined value of clearance *Cl* and effective plasma concentration C_e. During early preclinical screen such as during target validation/lead identification, ranking of compounds may be done by means of looking at the product of *in vitro* intrinsic clearance Cl_{int} and binding affinity K_d.

Why and how is scaling done? In discovery phases, scaling is done as part of the decision-making process in order to decide which compound might show the best pharmacokinetic and pharmacodynamic characteristics in man or if it is viable from a cost of goods perspective. It is primarily done to estimate a human dose, to predict the pharmacokinetic and pharmacodynamic parameters, and to simulate the concentration-time course, response-time course and therapeutic effect time course in man, or to demonstrate selectivity. In some cases such predictions may save a project drug from being stopped; for example, although the pharmacokinetic properties may not be optimal, the half-life of the response may still allow for a convenient dosing schedule in man. Thus, when the pharmacodynamic properties are scaled, not only are drug specific parameters such as EC_{50}/IC_{50} and E_{max}/I_{max} of interest, but so are the physiological or turnover parameters (k_{in} and k_{out}). The outcome variables or goals of applying biomarkers are to assess the effective concentration range and predictive plasma concentrations, and knowledge of the system properties such as secretion/synthesis and biomarker half-lives. The effective plasma concentration can be guided by, *e.g.*, an IC_{20}, IC_{50} or IC_{90} value, depending on the level of target occupancy or extent of biomarker response needed, as an initial goal. The system properties reveal how production and loss of a biomarker may be utilized for optimal drug treatment. For example, the short plasma half-life of omeprazole is not particularly informative of its rapid onset and extended duration of action. Nonetheless, the long turnover time, which is a system property of the proton pumps, combined with the irreversible action of the omeprazole metabolite, allows for a convenient low dose therapy.

Table 7.1 details some pharmacokinetic and pharmacodynamic data on felodipine, omeprazole and quetiapine. These are three typical examples of compounds with a pharmacokinetic profile that is far from optimal, and where good pharmacodynamic characteristics 'saved' the compound during development.

Table 7.1
Pharmacokinetic parameters in animals and man of felodipine, omeprazole and quetiapine.

Compound	Species	f_u (%)	F (%)	CL (mL·min^{-1})	$t_{1/2}$ (h)	Comment	Human dose (mg)
Felodipine	rat	1.5		108*			
	dog			1875*			
	man	<1	<15	5000*	5***	EC_{u50} <10 nM	2–10
Omeprazole	rat		<5	25	0.12		
	dog		15	150	1		
	man	3	60	4900	<1	$t_{1/2resp}$ 15–20 h	20–40
Quetiapine	rat	10	<10	41	0.35		
	dog	10	<10	450	1.5		
	man	17		1890	3.6	$t_{1/2resp}$ 1–2 wk	150–300**

* oral clearance, ** *b.i.d.*, *** effective half-life

When you read this chapter, try to think holistically, and include the following questions:

- When is scaling important, and when is it not?
- For what types of compounds is scaling less crucial?
- Why does scaling work for some compounds and not for others?
- What makes the drug that cannot be scaled different from those that can?
- What experimental data do we need to perform inter-species scaling with confidence?
- How and by what means may the uncertainties in predictions of *dose*, *Cl*, V_{ss} *etc.* be demonstrated?
- Under what circumstances can we best combine *in vivo* and *in vitro* data?

Scaling is, of course, not only pivotal for human therapeutic dose predictions, but also of great importance when predicting the safety margin in man. Whenever you embark on a scaling journey, ask yourself three questions: *What do you want to know?, How certain do you want to be?* and *What are you prepared to assume?* If you have a clear picture of the answers to those questions, scaling is seldom a problem.

7.2 What is allometry and allometric equations?

Energy consumption is of extreme importance in understanding interspecies variation in pharmacokinetics and pharmacodynamics, because it influences not only the amount of energy to be directed towards active storage and elimination of foreign chemicals (which includes drugs) but also indirectly, factors such as turnover rate and catabolism, that govern the pharmacological response. An increase in body mass is generally associated with a progressive reduction in the metabolic rate per unit mass, which in turn influences the turnover rate at the cellular level. Mitochondrial volume densities, membrane surface area densities, tissue weights, tissue specific and total tissue mitochondrial membrane surface areas are major determinants of energy turnover rates in small and large mammalian species. It has been shown that the mammalian liver varies in its 'leakiness' to sodium and potassium ions, and this 'leakiness' may at least in part explain the increased oxygen consumption. The larger specific area and the smaller the animal, the greater is the heat loss on a per kilogram body weight basis. The rat for example, has a greater number of miotchondria per gram liver than a human. This means that the smaller the animal the more energy is being turned over per unit of time in the resting state. If energy turnover is compared between small and large animals during exercise, it is more directly proportional to body weight since skeletal muscle energy turnover will be the main determinant of the total body heat generation. The number of mitocondria per gram of skeletal muscle tissue in small and large animals is similar. In this regard, the way in which oxygen consumption varies among species in the non-exercising state depending on size is illustrated in Figure 7.3.

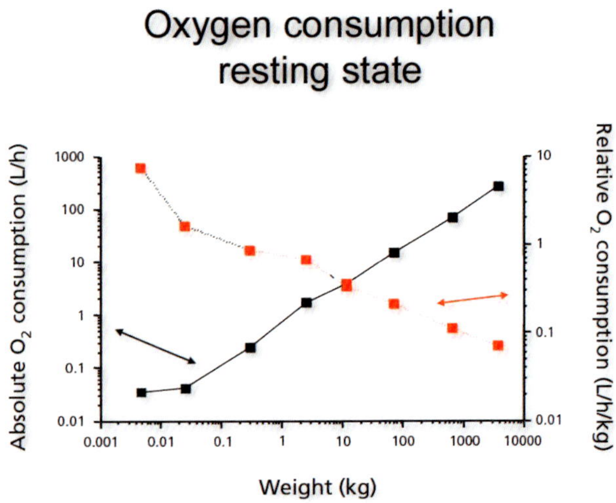

Figure 7.3
Data on absolute (filled black squares) and relative (filled red squares) oxygen consumption in different mammalian species ranging from the shrew (3-4 g) up to the elephant (4 tonnes) in the resting state.

Figure 7.4 illustrates how different rates of blood flow through the brain and liver and total cardiac output varies across different species, including humans. As can be seen there is a high regularity, with an allometric exponent of 0.8, and that the three variables increase in parallel across species.

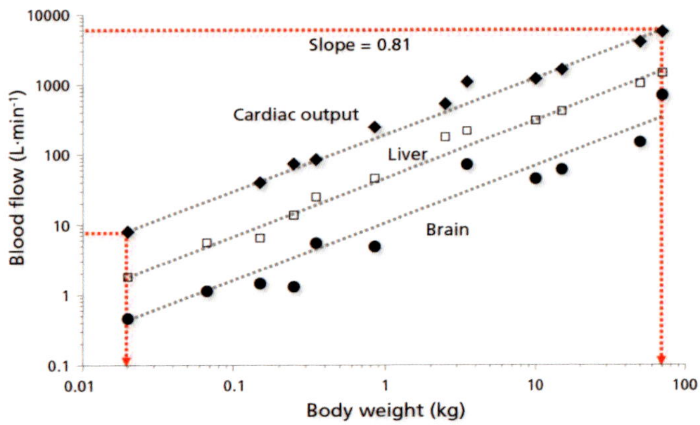

Figure 7.4
Log body weight (kg) *versus* log blood flow (L·min^{-1}) of brain and hepatic blood flow and total cardiac output with an exponent *b* of about 0.8. Data taken from Table 7.3.

The high rate of oxygen consumption of smaller animals in the resting state means that the oxygen supply, and hence blood flow, per gram tissue is about 100 times greater in the mouse (25 g), than in the elephant (4 tonnes). Other physiological variables such as heart rate, respiratory rate, food intake and turnover of drug are similarly affected. Why are pharmacokinetic parameters related to body weight? This is because numerous physiological parameters Y vary according to some mathematical function of body weight *BW*. The relationship between the physiological variable Y and size, such as body weight, is called allometry.

$$Y = a \cdot BW^{b} \tag{7:1}$$

For a given parameter, this allometric function appears to be reasonably uniform across a broad range of animal species. This regularity results in the ability to predict with a reasonably high degree of accuracy the parameters in a given species, including humans, based on the values found in other mammalian species. The present evidence suggests that this approach is particularly useful for substances that are eliminated primarily by physical transport processes (*e.g.*, biliary, renal, or pulmonary excretion), or when the disposition of the substance is largely independent of plasma protein binding. This phenomenon is considered more of a general biological regularity rather than a natural law since there are many examples of normal biological variation. These include human lifespan, the relative weight of the brain and the amount of oxygen consumed by the brain. However, for *qualitative* differences between different species in metabolism and/or excretion of chemicals, allometric scaling may fail. Some examples of this might be dissimilar metabolic routes, heterogeneity in phase I and phase II drug metabolism reactions, and differences in CYP450 isozyme makeup.

It has also been shown that processes related to the metabolic rate of an organism, such as clearance *Cl*, is often proportional to the 0.75 power of body weight

$$Cl = a \cdot BW^{0.75} \tag{7:2}$$

where *a* is a constant of appropriate units and *b* is the allometric exponent. When plotted on a log-log scale, this is a straight line with intercept log(*a*) and slope *b*.

$$\log(Cl) = \log(a) + b \cdot \log(BW) \tag{7:3}$$

Figure 7.5 illustrates a pharmacokinetic experiment where methadone was given as an intravenous injection to mice, rats, monkeys, dogs and humans. It is clear that the terminal half-life gets longer as the body weight increases. The half-life is less than 15 min in mice whereas it is in the range of 15–75 h in man.

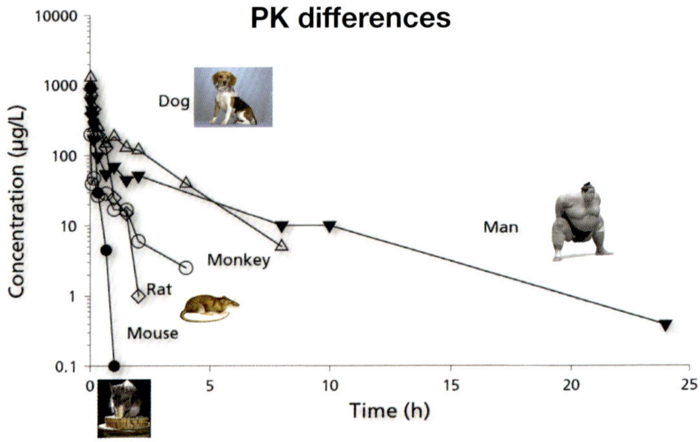

Figure 7.5
Mean concentration versus time data of methadone in mice, rats, monkeys, dogs and humans following an intravenous injection dose. Note how the terminal half-life gets longer with an increased size of the animal.

When extrapolating within and across species, the use of regression analyses across *body weight* has been recommended as the most accurate methodology. This is because of difficulties inherent in measuring body surface parameters as opposed to body weight at the individual level. As a result, body surface parameters usually require the use of a reference criterion. Several theoretical possibilities have been suggested as to why surface area might have a poorer predictability across species than body weight. The surface area of living organisms is difficult to determine and verify. For example, various species increase their coat thickness, or subcutaneous fat at different times of the year and in response to environmental factors, such changes alter their effective surface area to help control body temperature. While surface area changes with the 2/3 power of body weight, this is only true for geometrically similar bodies of constant specific gravity. It is a gross over simplification to accept that surface area determines the metabolic rate. It has been suggested that it is more rational to use BW^b instead of $BW^{2/3}$. Looking at the distribution of the allometric exponent b across about 100 compounds, displays a wide variability from about 0.3 to 1.2, and almost normal distribution with a mean of 0.78 (Figure 7.6).

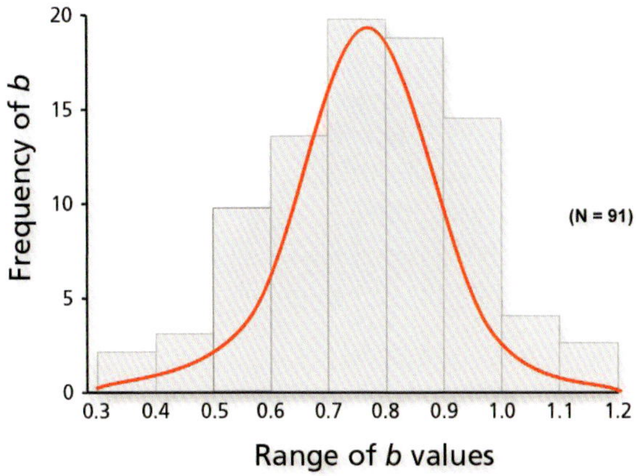

Figure 7.6
Histogram of different values of the allometric exponent *b* collected from 91 different compounds. In spite of a wide range from about 0.3 to 1.2, it is centered on 0.7-0.8 which agrees well with the theory of energy turnover. The red line displays an approximate normal distribution.

The calculation of body weight is direct, simple and often performed with data from individual animals other than the standard reference mouse, gerbil, hamster, rat, marmoset, guinea pig, rabbit, rhesus monkey, dog, minipig, sheep and man (see Tables 7.4 and 7.5). The body weight rule offers the scientist a very powerful tool with which to make reasonable interspecies predictions. Note that there are compounds for which this general rule fails and others for which it is applicable. In the real world, it is highly recommended that data be available from four species under similar, standardized conditions when performing regression analysis. Examples of the allometric relationship between some physiological variables and body weight are shown schematically in Figure 7.7 using Equation 7:3.

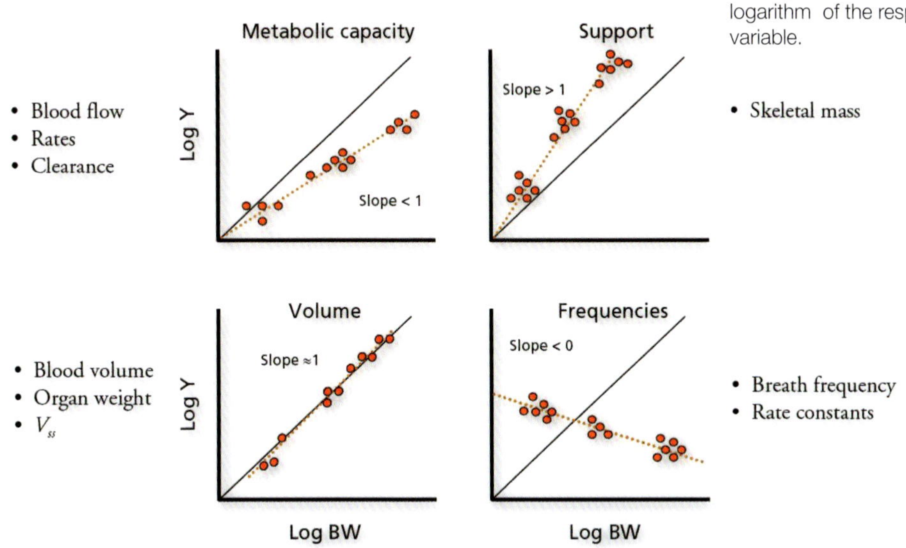

Figure 7.7
Allometric plots with different exponents. Solid black lines illustrate theoretical relationships with an exponent of unity. **Upper left plot** shows how the metabolic rate increases with body weight ($b \sim 0.75$). **Upper right plot** illustrates how support (e.g., cross-sectional area of skeletal bones) of mammals increases out of proportion to body weight ($b \sim 1.08$). **Lower left plot** shows how some variables, like blood volume, increase proportionately with body weight ($b \sim 1$). **Lower right plot** shows how e.g., heart rate decreases with increase in body weight ($b \sim -0.25$). Log *Y* and log BW denote the natural logarithm of the respective variable.

When plotted on a log-log scale, the relationship is a straight line with intercept log a and slope b. The physical nature of parameters a and b is different. Commonly, it is said that a is drug dependent and b is not. Rather, the latter is dependent on the type of physiological variable being analyzed *e.g.*, *Cl*, *V* or $t_{1/2}$. The impact of two different exponents on the relationship between Y and body weight in Equations 7:1 and 7:3 is demonstrated graphically in Figure 7.8.

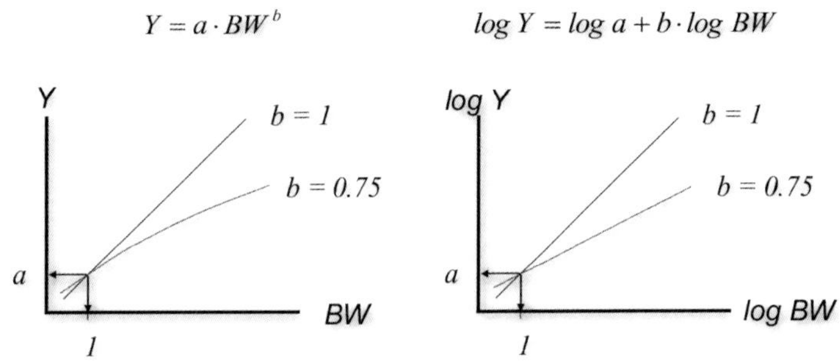

Figure 7.8
The left hand figure demonstrates on a linear scale the relationship between a physiological variable *Y* and body weight for two different exponents acting on body weight *BW*. When *b* is equal to 1 (unity), a direct proportionality is present between *Y* and body weight. When *b* is equal to 0.75, a curvilinear relationship is present. Both relationships have the same value when body weight is equal to 1 (unity). The right hand figure shows the same relationships plotted on a log-log scale where log denotes the natural logarithm.

As stated above, the constants a and b give the intercept at a body weight of unity (1) and the slope in a log-log plot, respectively (Figure 7.8). It is unrealistic to expect that a regression line (or an equation) will indicate precisely what the specific metabolic rate or other variable in question will be for an animal of a given body weight. Rather, what the equation will provide is the expected mean value for a typical mammal of the given size. Animals will always deviate more or less from this idealized norm, thus a good correlation does not necessarily imply a good prediction.

The choice of species (body weight) will also have an impact on the precision of the predicted parameter. The upper panel in Figure 7.9 illustrates schematically how the precision in, for example, *Cl* decreases when one tries to predict *Cl* in man from three different species with the same uncertainty in the allometric exponent b. Using that reasoning, the closer the animal is to man with respect to body weight and clearance mechanisms, the smaller the uncertainty in the prediction will be. The lower panel demonstrates how a downward shift in the animal's *Cl* will affect the final point estimate independently of the body weight of that particular animal. Experience suggests that the exponent b for *Cl* can vary from about 0.3 to 1.2 depending on, for example, the quality of data, the number of species included, nonlinearities and lack of correction for species dependent plasma protein binding (Figure 7.6).

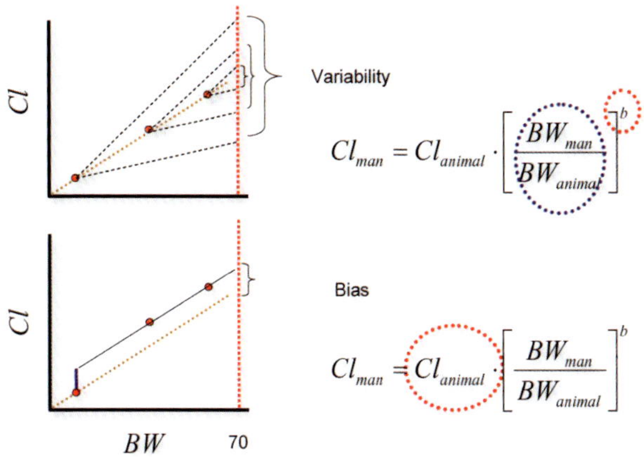

Figure 7.9
Upper plot schematically illustrates how the distance in body weight away from 70 kg increases the variability in estimated human clearance. The lower plot shows how a vertical shift in animal clearance affects predicted human clearance independently of body weight. The inset allometric equation for clearance demonstrates the three major sources of variability: Measured animal clearance, body weight ratio between man and animal, and scaling exponent *b*. This also shows the advantage of using more than 3 species in allometric scaling predictions to (partly) offset any potential 'fulcrum' effect of the species in either end of the prediction curve.

The allometric relationship between an animal clearance and human clearance is given by Equation 7:4

$$Cl_{man} = Cl_{animal} \cdot \left(\frac{BW_{man}}{BW_{animal}}\right)^b \tag{7:4}$$

When clearance in man is predicted from different degrees of variability in Cl_{animal}, body weight ratio BW_{man}-*to*-BW_{animal} and the allometric exponent *b*, it will result in a predicted skewed distribution of Cl_{man}. The reason for this is the exponential term. This is also the case for human dose predictions allometrically scaled from animals, provided the same areas under the plasma concentration-time curves *AUC* from zero to infinity.

$$\begin{cases} Cl_{man} = \dfrac{Dose_{man}}{AUC_0^\infty} \\ Cl_{animal} = \dfrac{Dose_{animal}}{AUC_0^\infty} \end{cases} \tag{7:5}$$

Replacing Cl_{man} and Cl_{animal} by $Dose_{man}/AUC$ and $Dose_{animal}/AUC$, respectively in Equation 7:4, then the *AUCs* cancel out and Equation 7:4 can be rewritten as

$$Dose_{man} = Dose_{animal} \cdot \left(\frac{BW_{man}}{BW_{animal}}\right)^b \tag{7:6}$$

This skewed distribution of both clearance and *Dose* on a linear scale may exhibit a more uniform distribution on a logarithmic scale. The allometric exponent *b* can also vary within and between different species and across data pooled from several species (Figure 7.10).

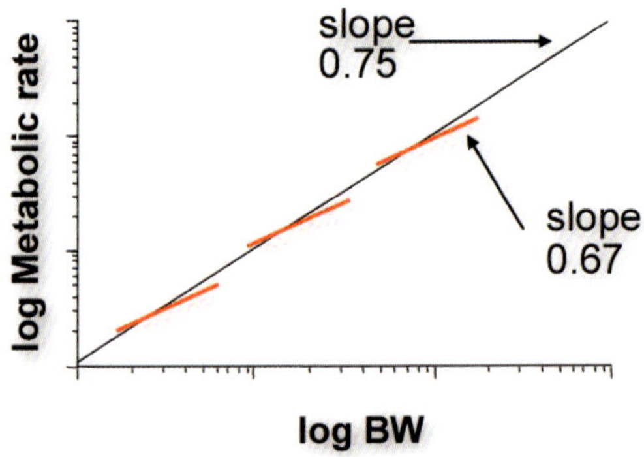

Figure 7.10
Schematic illustration of how the allometric exponent *b* might differ depending on whether data are obtained within a species or among several species. The 0.67 exponent may well be a correct descriptor of how the metabolic rate varies within a species, although among several species, an exponent of 0.75 (3/4) emerges to be a more commonly used descriptor.

Figure 7.11 shows schematically how the allometric constant *a* is obtained graphically from clearance data in the rats, monkey, dog and man. The *a* parameter is the intercept of the Y-axis for a body weight equal to 1 kg.

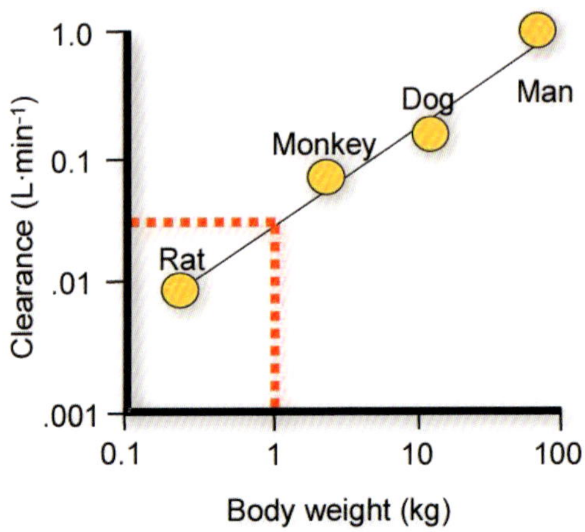

Figure 7.11
Log body weight (kg) *versus* log clearance ($L \cdot min^{-1}$) of compound X with an exponent *b* of about 0.75 and an intercept *a* of about 0.03 $L \cdot min^{-1}$.

An allometric model taking total body weight into account has also been derived for volume of distribution. The exponent *d* of the volume term is often close to unity

$$V = c \cdot BW^{d} \tag{7:7}$$

where c is a constant of appropriate units. When plotted on a log-log scale, this is a straight line with intercept log(c) and slope d.

$$\log(V) = \log(c) + d \cdot \log(BW) \tag{7:8}$$

Substitution of the volume and clearance in a mono-exponential model by Equations 7:2 and 7:7 gives

$$C = \frac{Dose_{iv}}{V} \cdot e^{-\frac{Cl}{V} \cdot t} = \frac{Dose_{iv}}{c \cdot BW^{d}} \cdot e^{-\frac{a \cdot BW^{b}}{c \cdot BW^{d}} \cdot t} \tag{7:9}$$

The application of Equation 7:9 when simultaneously fitting plasma concentration-time data of mice, rats and humans are shown in Figure 7.12 (left). The analysis is also extended to construction of a Dedrick plot (Figure 7.12, right). A Dedrick plot has species-independent concentration and time axes. The concentration axis is dose and body weight normalized.

7.3 The elementary Dedrick plot

This Section discusses the application of how to derive species-independent concentration and time scales to the study of inter-species comparisons of pharmacokinetics. Plasma concentration-time data of methadone were obtained from mouse, rat and man after receiving an intravenous bolus of 25, 500 and 100,000 μg, respectively (Figure 7.12). The corresponding body weights were 23 g (mouse), 250 g (rat) and 70 kg (man), covering a 3,000-fold weight range. An instantaneous distribution is supposed to occur following drug administration. A plot of dose and body weight-normalized (scaled) plasma concentrations ($C \cdot D^{-1} \cdot BW^{1}$) *versus* body weight-normalized times ($t \cdot BW^{b-1}$, which is time normalized or scaled by clearance) shows a straight line and allows superposition of the curves. This is called the elementary Dedrick plot. The X axis represents a pharmacokinetic unit of time during which the different species clear the same volume of plasma per kilogram of body weight (so-called kallynochron).

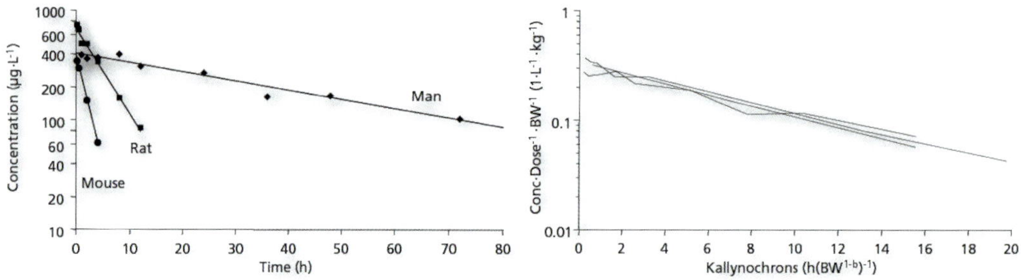

Figure 7.12
Observed (symbols) and predicted (lines) plasma concentrations in mouse, rat, and man following an intravenous bolus dose of methadone (left figure). The right figure shows concentrations normalized for dose and body weight, *versus* time normalized for clearance.

7.4 Examples of allometric plots

This Section covers some further examples of allometric plots in order to motivate the use of allometry and how to interpret results. As has been shown previously, an allometric plot is the body weight *versus* clearance or volume of distribution plotted in a double-logarithmic diagram. When the allometric exponent b is less than 1, a doubling of body weight produces less than a 2-fold increase in the dependent variable. Clearances frequently fall into this category. Consequently, with larger species, doses (in mg·kg^{-1}) generally get smaller. When the exponent is 1, a constant ratio exists between variables. Blood volume in terrestrial mammals scales with an exponent of 0.99 and tends to be about 7 % of body mass, regardless of size. Skeletal mass in terrestrial mammals scales with an exponent of 1.09. Thus, larger mammals have proportionately more of their body mass dedicated to bone than smaller ones. For example, the skeleton of a shrew is about 5 % of body mass, whereas that of an elephant is 27 % (Fig. 7.7, Table 7.2).

Table 7.2
Impact of a doubling of body weight on physiological variables with different exponents.

Exponent	Factor for change in dependent variable as BW doubles	Examples of dependent variables
0.0	0.0	Hematocrit
0.25	1.19	Lifespan, disposition half-life, breath time, heart beat time
0.67	1.59	Body surface area
0.75	1.68	Metabolic rate, clearance
1.0	2.0	Blood volume, volume of distribution
1.08	2.11	Skeletal mass

Two real life cases are shown when human predicted clearance is relatively well-defined (Figure 7.13) and poorly defined (Figure 7.14).

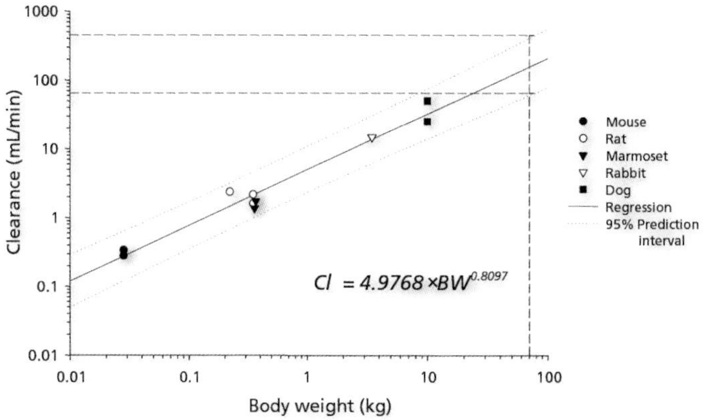

Figure 7.13
Log body weight (kg) *versus* log clearance (L·min^{-1}) of compound X with an exponent *b* of about 0.8 and an intercept a of about 5 L·min^{-1}. Data are obtained from mouse, rat, marmoset, rabbit and dog.

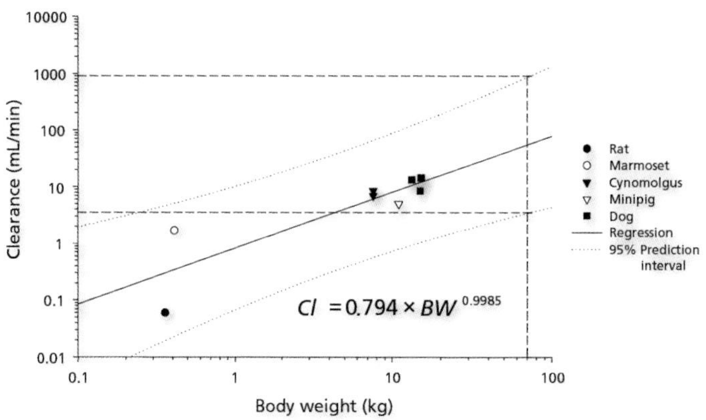

Figure 7.14
Log body weight (kg) *versus* log clearance (L·min^{-1}) of compound X with an exponent *b* of about 1.0 and an intercept a of about 0.8 L·min^{-1}. Data are obtained from rat, marmoset, cynomolgus monkey , minipig and dog.

Figure 7.15 demonstrates a common difficulty when predicting a reasonable first dose in man based on allometric data. The predicted dose range is large and this does not help to design an appropriate tolerability study in man. The large dose range may also jeopardize reasonable estimation of the cost of goods for compound production, ranging from a large tablet (500 mg) to a hockey puck size of tablet (3,000 mg). Both the marmoset and the rat will have a substantial leverage effect on the slope of the curve and thus prediction to man, and it is difficult to judge which one of them is the best predictor in this particular case. One might also want to collect kinetic information from the less than 100 g (*e.g.*, mouse) and 1–5 kg (*e.g.*, rabbit) body weight ranges.

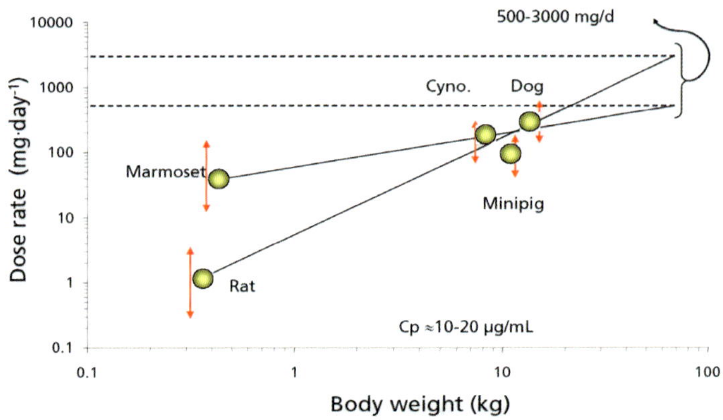

Figure 7.15
Log body weight (kg) *versus* log dose rate (mg·day^{-1}) of compound X with an exponent *b* of about 1.0 and an intercept *a* of about 0.8 L·min^{-1}. Data are obtained from rat, marmoset, cynomolgus monkey (Cyno.), minipig and dog. The double arrows show the relative impact on leverage.

However, bear in mind that the crucial questions to ask are what costs are saved in the practical situation in the clinic from fine-tuning scaled preclinical data, and what therapeutic areas would benefit from more precisely scaled parameters such as *Cl* and daily dose. When predicting the dose the effective concentration range also needs to be incorporated. Sometimes the effective concentration range may be large or even unknown from *in vivo,* and then *in vitro* data have to be used for direction.

The development of so-called µ-dosing

7.5 Time scales differ between different species

Physiological time is defined as a species dependent unit of chronological time required for the completion of a species independent physiological event. For example, life expectancy is 14 years in the dog and 98 years in man. This means that the dog consumes 7.14 % of its life every year, whereas man consumes 7.14 % of his life every 7 years. Thus, one year for the dog and 7 years for man correspond to 7.14 % of the lifetime of these species (Boxenbaum [1982]). Boxenbaum concluded that one year in the dog and seven years in the man are equivalent physiological times necessary to produce a species independent physiological event, *i.e.*, living 7.14 % of a lifetime. Other examples of physiological times are breath time and heart beat time. These differ considerably between mammalian species as follows

$$Breath\ time = 0.169 \cdot BW^{0.28} \tag{7:10}$$

$$Heart\ beat\ time = 0.0428 \cdot BW^{0.28} \tag{7:11}$$

$$\frac{Breath\ time}{Heart\ beat\ time} = \frac{0.169 \cdot BW^{0.28}}{0.0428 \cdot BW^{0.28}} \approx 4 \tag{7:12}$$

Dividing breath time by heart beat time gives a value of about 4. In other words, all mammals have 4 heart beats for each respiratory cycle. Lifetime or lifespan scales with an exponential factor of 0.28. This indicates that most mammals have nearly the same number of breaths per lifetime and the same number of heart beats per lifetime even though their lifespans differ. This also relates to pharmacokinetics. Mammals tend to clear drugs at a similar pace, measured by their own internal clocks. The relationship between the half-life of a drug in plasma and clearance and volume

$$t_{1/2} = log(2) \cdot \frac{V}{Cl} = log(2) \cdot \frac{c \cdot BW_i^d}{a \cdot BW_i^b}$$
$$= constant \cdot BW^{d-b} \propto BW^{0.25} \tag{7:13}$$

where BW^{d-b} is approximately $BW^{0.25}$ since exponent for volume d is close to 1 and exponent for clearance b is approximately 0.75 with a range of 0.3 to 1.2. The maximum lifespan potential *MLP* is the maximum documented longevity for a species. Sacher [1959] fitted the following equation to data of 63 mammalian species

$$MLP = 10.839 \cdot W^{0.636} \cdot BW^{-0.225} \tag{7:14}$$

BW and *W* are the body weight and brain weight, respectively. Using the equation above, the ratio of MLP_{man} to MLP_{dog} was a factor of 4.7 (Table 7.3). Comparing MLP_{man} to MLP_{rat} the factor is 20. Taking only *MLP* into account, the exposure to a concentration of unity in the rat during 1 h corresponds to a 20 h period in man at the same concentration level, even though the absolute *AUC* will be 20 times greater in man.

Species	Body Weight BW (g)	Brain Weight W (g)	% W/BW	MLP (years)
Mouse	20	0.36	1.8	2.9
Gerbil	67	0.98	1.46	4.2
Rat	250	1.88	0.75	4.7
Marmoset	350	8.0	2.3	10.9
Dog	10,000	80	0.80	22.1
Rhesus Monkey	5,000	90	1.8	27.9
Man	70,000	1,500	2.14	92.2

Table 7.3
Maximum lifetime potential of different species (Data from Tables 7.4 and 7.5)

By means of information about clearance and dose in one species, one may be able to predict the appropriate exposure or dose level in another animal or in man. The body weight rule offers the investigator a very powerful tool with which to obtain general guidelines for making reasonable inter-species predictions. Furthermore, preclinical evaluation and analysis may reduce the frequency and severity of unexpected toxic events during single dose phase I testing, because such data enable predictions to be made about dose and exposure. For example, if one wishes to establish similar unbound exposure in terms of the area under the unbound plasma concentration-time curve within a dosing interval AUC_u in the laboratory rat and man

$$AUC_{u,rat} = AUC_{u,man} = \frac{Dose_{man}}{Cl_{u,man}} = \frac{Dose_{rat}}{Cl_{u,rat}} \quad (7:15)$$

Clearance can be written as a function of body weight

$$Cl_{u\,i} = a \cdot BW_i^b \quad (7:16)$$

and then inserted into Equation 7:15 gives

$$\frac{Dose_{man}}{a \cdot BW_{man}^b} = \frac{Dose_{rat}}{a \cdot BW_{rat}^b} \quad (7:17)$$

By rearranging Equation 7:17, one obtains for dose predictions

$$Dose_{man} = Dose_{rat} \cdot \left[\frac{BW_{man}}{BW_{rat}}\right]^b \quad (7:18)$$

Remember though, that Equation 7:18 is based on the assumption of equal sizes of unbound areas in man and the beast, although the shapes of the areas may differ substantially.

7.6 Physiological variables of 11 animal species and man

It is our firm belief that allometric scaling works for the majority of compounds, provided the data are well behaved and with good precision. There might be some instances when scaling fails for a number of reasons. These can include binding differences across species, qualitative metabolic differences such as different isozymes involved, polymorphism, complex concentration-time profiles (e.g., enterohepatic recirculation), non-constant volume ratios (e.g., V_c/V_t variable across species), nonlinear processes, qualitative differences in extrahepatic elimination and different routes of elimination. Tables 7.4 and 7.5 contain a compilation of organ weights and perfusion rates for a number of species. The total blood volume for the various mammalian species in Table 7.5 can be approximated to 7 % of the total body weight. In other words, man will have a blood volume of about 5 L and the 250 gram rat about 20 mL. Keep in mind that the values of tissue/organ sizes and blood flows in Tables 7.4 and 7.5 may vary substantially between and within different individual animals and strains due to the actual physiological state and the experimental method applied (for example, the injected microspheres technique), anaesthetics used, the surgical approach, body composition and stress induced. Consequently, it is not uncommon to have a biological variation of greater than 30 %. The impact of this variation on the prediction of exposure or kinetics can only be foreseen by means of Monte-Carlo simulations.

Table 7.4
Organ weights (g) and volumes (mL) of CSF, blood, plasma, total body water TBW, intracellular fluid, and extracellular fluid (mL) in mouse, gerbil, hamster, rat, marmoset, guinea pig, rabbit, rhesus monkey, beagle dog, minipig, sheep and man

	Mouse (0.02kg)	Gerbil (0.067kg)	Hamster (0.15kg)	Rat (0.25kg)	Marmoset (0.35kg)	Guinea pig (0.85kg)	Rabbit (2.5kg)	Rhesus monkey (5kg)	Dog (10kg)	Minipig (15kg)	Sheep (50kg)	Human (70kg)
Brain	0.36	0.98	1.4	1.8	8.0	4.4	14	90	80	120	400	1500
Liver	1.75	2.1	6.9	10.0	19.0	38	77	150	320	480	1600	1500
Kidneys	0.32	0.55	2.0	2.0	2.4	8.8	13	25	50	75	250	310
Heart	0.08	0.25	0.6	1.0	2.4	3.6	5	18.5	80	120	400	330
Spleen	0.1	0.049	0.54	0.75	0.6	2.1	1	8	25	38	125	180
Adrenals	0.004	0.038	-	0.05	0.1	0.90	0.5	12	1	1.5	10	14
Lung	0.12	0.32	0.74	1.5	1.9	10.1	18	33	100	150	1000	1000
Muscle	10.	27.	68.7	120	144	320*	1350	2500	4600	6900	27700	35000
GI tract	1.5	2.8	19.2	6.3	7.6	43	120	230	190	285	3950	2100
Fat	-	4.6	17.	17.	20	-	-	-	-		-	10000
Marrow	0.6	-	-	-	-	-	40	135	110	165	-	1400
Skin	-	-	-	-	-	167	-	-	-		-	
Bone	-	-	-	-	-	-	-	-	1180	1770	6930	10000
CSF mL	0.035	0.04	0.15	0.15	0.18	-	2.3	-	-	-	-	100
Blood mL	1.4–1.6	5.4	10	16	25	50	140	280	850	975	3600	5000
Plasma mL	1			7.8		-		224	515	-	-	3000
TBW mL	14.5			167			1790	3465	6035			42000
Intracell.				93			1165	2425	3276			23800
Extracell.				74			625	1040	2760			18200

Minipig data scaled from dog by means of a scaling exponent of 1 (unity).

Table 7.5
Organ blood flows (mL·min^{-1}), urine flow (mL·day^{-1}), bile flow, CSF flow and GFR in mouse, gerbil, hamster, rat, marmoset, guinea pig, rabbit, rhesus monkey, beagle dog, minipig, sheep and man Gabrielsson and Weiner (2010)

Blood flows (mL·min^{-1})	Mouse (0.02kg)	Gerbil (0.067kg)	Hamster (0.15kg)	Rat (0.25kg)	Marmoset (0.35kg)	Guinea pig (0.85kg)	Rabbit (2.5kg)	Rhesus monkey (5kg)	Dog (10kg)	Minipig (15kg)	Sheep (50kg)	Human (70kg)
Brain	0.46	1.13	1.45	1.3	5.5	4.9	-	72	45	61	150	700
Liver	1.8	5.6	6.5	13.8	25	46	177	218	309	419	1033	1450
Kidneys	1.3	2.82	6.2	9.2	10.5	36	80	138	216	293	722	1240
Heart	0.28	1.12	1.4	3.9	4.47	9.9	16	60	54	73	181	240
Spleen	0.09	0.07	0.25	0.63	0.72	4.6	9	21	25	34	84	77
Gut	1.5	3.58	5.3	7.5	13.6	39	111	125	216	293	722	1100
Muscle	0.91	0.1	4.5	7.5	8.6	-	155	90	250	339	836	750
B. Marrow	0.37	-	-	-	-	-	24	51	37	50	-	270
Adipose tiss.	-	-	0.77	0.4	0.46	-	32	20	35	47	117	260
Skin	0.41	-	-	5.8	6.6	29	-	54	100	136	334	300
Hepatic artery	0.35	0.30	1.2	4.0	7.3	7.4	37	51	79	107	264	300
Portal vein	1.45	5.3	5.3	9.8	11.2	-	140	167	230	312	769	1150
Cardiac output	8.0	-	40	74.0	85	248	530	1086	1200	1626	4010	5600
Urine flow (mL·day^{-1})	1.0	-	-	50.0	57	<100	150	375	300	407	1000	1400
Bile flow (mL·day^{-1})	2.0	-	-	22.5	25.8	-	300	125	120	163*	400	350
CSF flow (mL·h^{-1})	0.018	-	-	0.18	0.76	-	0.6	2.5	-	-	-	21
GFR (mL·min^{-1})	0.28	-	-	1.31	1.5	-	7.8	10.4	61.3	83	205	125

7.7 Allometric scaling of turnover parameters

Another major application of allometry to inter-species prediction is with pharmacodynamic turnover data. Here, turnover is composed of the turnover rate and fractional turnover rate. This Section shows how information about the endogenous turnover of compounds such as proteins, peptides and hormones, can be scaled across different species. It also shows how some of these are related to energy turnover or metabolic rates. Figure 7.16 shows the metabolic rate and protein turnover rate in a number of mammalian species across four orders of magnitude in body weight.

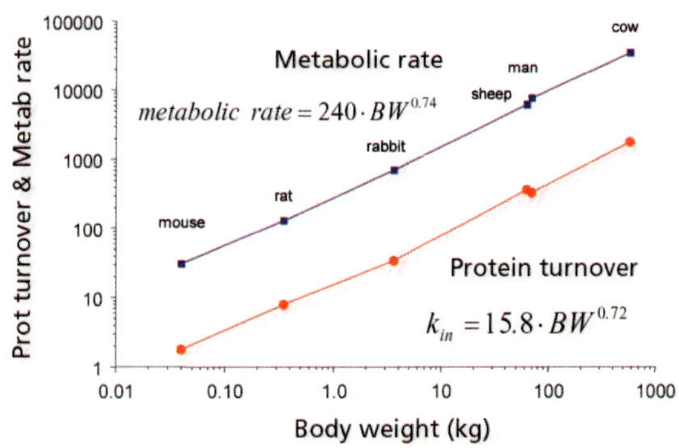

Figure 7.16
Data on metabolic rates and total protein turnover rates of various named species.

Figure 7.17 details the half-lives of albumin and α_1-acid glycoprotein in several species including man. Note the parallel shift between the two lines. A half-life of albumin in any species is on average 2.5 times longer than the half-life of α_1-acid glycoprotein within the same species.

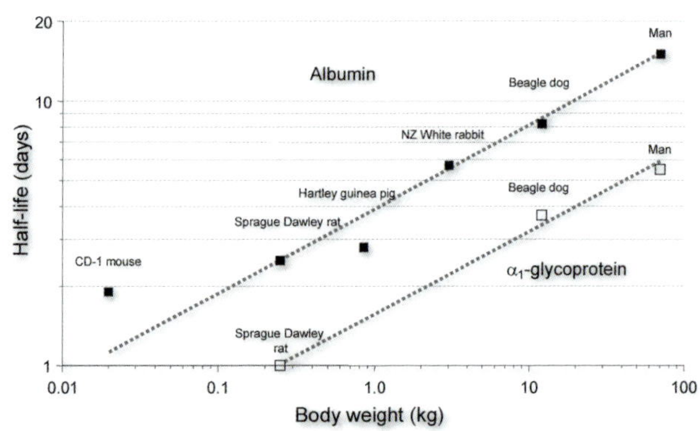

Figure 7.17
Data on half-lives of albumin and α_1-glycoprotein AAG of various named species. Note how perfectly both protein half-lives scale across a wide mammalian body weight range. The albumin half-life is about 2.5 times longer than that of AAG.

These examples on albumin, α_1-acid glycoprotein (Figure 7.17) and total protein turnover (Figure 7.16) show how some large molecules and metabolic processes scale in a predictive manner across different mammalian species in the resting state.

Let us therefore look at some basic principles of allometry and see how they apply to turnover. Most rate processes like blood flows, clearances, secretion rates, and synthesis rates require energy and scale generally proportionally to body weight raised to the power b of 0.3–1.2, but it is common to approximate b to 0.7–0.8. The general equation for turnover rate k_{in} is

$$k_{in\,man} = k_{in\,beast} \cdot \left(\frac{BW_{man}}{BW_{beast}} \right)^b \qquad (7:19)$$

A simulated time course of response for rat and man is illustrated in Figure 7.18. It is assumed that potency and intrinsic activity are the same for the drug in the two species and that the only difference is due to $t_{1/2}$ response ($t_{1/2kout}$).

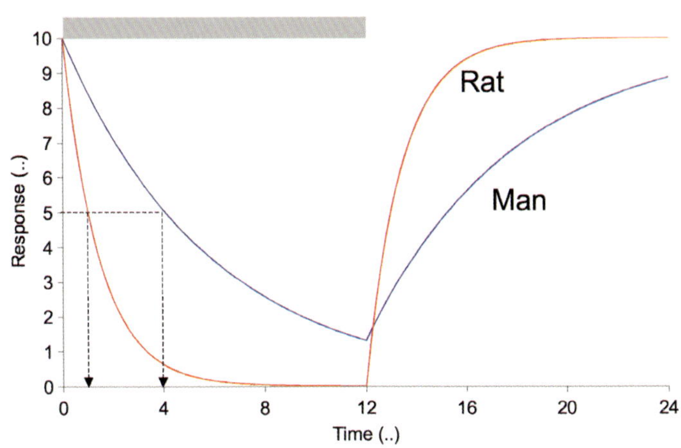

Figure 7.18
Response-time course in rat (red line) and man (blue line) following the same period of exposure (horizontal grey bar. Time is arbitrary units. The time to reach 50 % of maximal response is approximately one time unit in the rat and four time units in man.

7.8 General conclusions

As a general rule extrapolations across different species are valid. There are, however, situations where inter-species extrapolations may lead to errors. Rather than reporting every case when scaling is valid, it is more informative to document what makes the unscalable drug different. The compilation of different sources of variability in Table 7.7 may be used as a list of points to consider when scaling. In addition to the uncertainty in the predicted unbound clearance, one should also consider the variability in the unbound target concentration when predicting clinical doses.

Table 7.7
Points to consider when scaling

Critical factor	Comments
Strain	Different strains, *e.g.*, Wistar and Sprague Dawley rats may exhibit different metabolic capacities
Gender	Hormonal, anatomical, and physiological differences
Capacity dependent elimination	Deviations from first-order kinetics (use V_{max} and K_m when scaling)
Size and age	Use explicit weight
Insufficient technical detail	Sample times, duration of sampling, number of subjects, quality of underlying data that the scaling exercise is based on
Acute *versus* multiple dosing	Impact of induction/inhibition
Protein/tissue binding	Nonlinear binding, inter-species differences in binding, disease state, malnutrition
Disease state(s)	Metabolic and physiological status
Routes of administration	Extravascular routes contain absorption steps with increased variability
Elimination pathways	Qualitative and quantitative differences between species
Source(s) of variability	Effective concentration range, within species variability in body weight and kinetics, variability in allometric constants and exponents, skewed predicted dose range

In the following sections, two different Case scenarios are presented, to further illustrate and make the reader more familiar with the general principles of allometric scaling and its applications in drug development.

7.9 Dose nomogram and scaling of pharmacodynamic parameters

The prediction of human dose has so far been focused on allometric extrapolation of pharmacokinetic parameters and how to scale clearance across species. However, equally important is the assessment of the pharmacodynamic parameters, such as potency (e.g., EC_{50}) across different species and the assessment of safety margin. This can be illustrated with a practical example. The goal of this Case Study is to demonstrate one approach for estimation of a *human daily dose range* of a growth stimulating compound. The response data will be used for selection of an appropriate therapeutic (effective) concentration interval and the kinetic data then used for allometric extrapolation of laboratory animal data to man (Figures 7.19–7.20).

The biomarker in this example is a hormonal response (release of a growth hormone and covers an interval between 50 (approximately baseline) and 120 units (Figure 7.20). This Case Study covers four steps.

Step 1 deals with making an allometric plot of clearance of various species versus body weight.

Step 2 involves the analysis of total and unbound plasma concentration-response plots.

Step 3 combines the results of Step 1 and 2. In **Step 3** the human daily dose is predicted prior to

Step 4 where a dose nomogram is constructed.

Step 1

Analyze the data in Table 7.8 and estimate the unbound human oral clearance Cl_u/F. Allometry gives an approximate human unbound oral clearance Cl_u/F of 5–6 L·min^{-1}.

Table 7.8
Compilation of physiological data including body weight, free fraction, hepatic blood flow, systemic clearance, hepatic extraction, availability across the liver, oral clearance and free oral clearance of compound X for mouse, rat and dog.

Species	Body weight (kg)	f_u	Q_H (L·min^{-1})	Cl_{iv} (L·min^{-1})	E_H	F_H	Cl/F (L·min^{-1})	Cl_u/F (L·min^{-1})
Mouse	0.025	0.01	0.0018	0.0005	0.28	0.72	0.00069	0.069
Rat	0.3	0.03	0.015	0.003	0.20	0.80	0.004	0.13
Dog	15	0.1	0.31	0.15	0.48	0.52	0.291	2.91

Figure 7.19
Double-logarithmic plot of body weight (kg) *versus* unbound oral clearance in the mouse, rat and dog. Data obtained from Table 7.8.

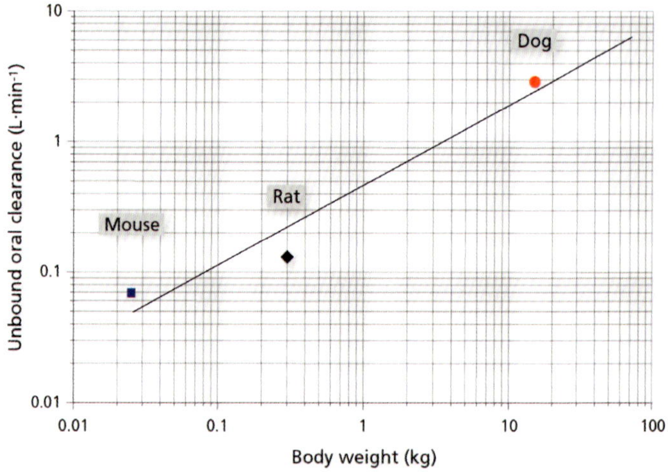

Total concentration-response data of a biomarker were obtained after giving a new experimental compound to mice, rats and dogs at several dose levels (Figure 7.20, left). It seems that the test compound has different potencies in the three species, with highest potency (EC_{50} lowest) in dogs and lowest in mice (EC_{50} highest). However, this impression is only imaginary due to the large species-dependent differences in the plasma protein binding of test compound. Indeed when the total concentration-response data are transformed to unbound concentration-response relationship, the data superimpose (Figure 7.20, right).

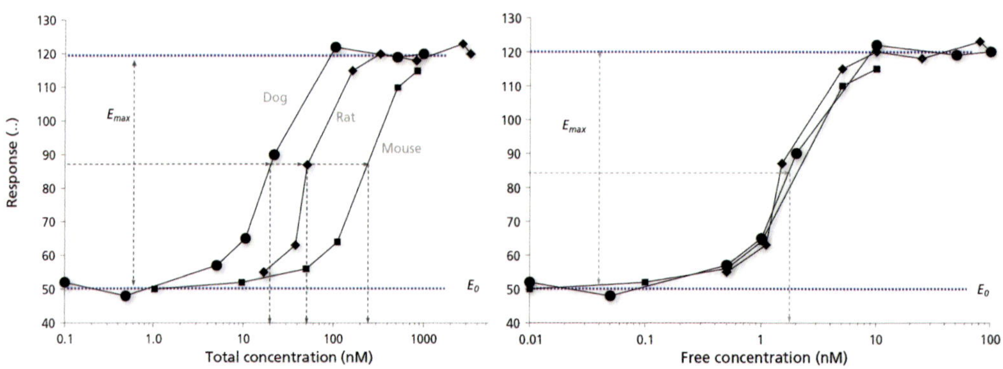

Figure 7.20
Semi-logarithmic plot of total plasma concentration versus a biomarker response (left) and unbound plasma concentration *versus* biomarker response (right) in dogs, rats and mice at several dose levels of test compound X.

A baseline response E_0 of about 50 response units is measured when no drug is present. The maximum observed response is seen at about 120 units. Consequently, the maximum drug induced response E_{max} corresponds to 70 (120–50) units. The plasma concentration at 50 % of E_{max} is called the potency EC_{50} (Figure 7.20, right). Total concentration-response data are shown in Table 7.10. Note the large concentration range.

Step 2a

Analyze the unbound concentration-response data by converting total concentrations in Table 7.10 to the unbound plasma concentrations by means of the free fractions of compound X for mouse (0.01), rat (0.03) and dog (0.1). and estimate the primary pharmacodynamic parameters (Figure 7.21, right).

Figure 7.21
Semi-logarithmic plot of total plasma concentration *versus* a biomarker response (left) and unbound plasma concentration versus biomarker response (right) in dogs, rats and mice at several dose levels of test compound X.

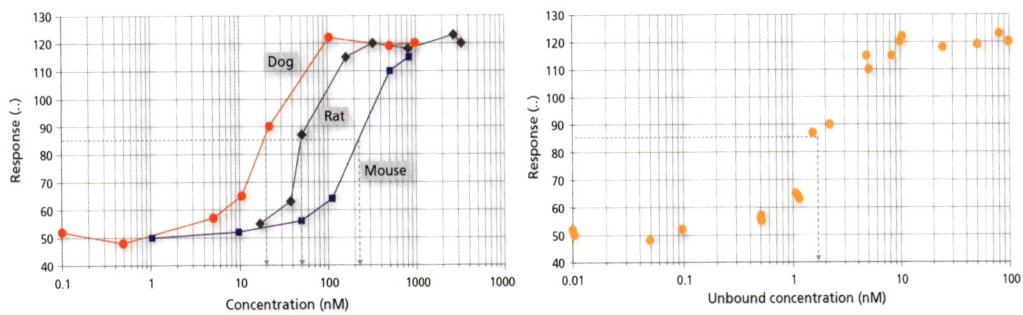

Nonlinear regression of unbound concentration-response data (which superimpose for all species) by means of the *Hill-equation* gives

$$E = E_0 + \frac{E_{\max} \cdot C_u^n}{EC_{50}^n + C_u^n} = 50 + \frac{70 \cdot C_u^{2.2}}{1.7^{2.2} + C_u^{2.2}} \qquad (7:20)$$

where E_0, EC_{u50}, E_{max} and n are the baseline value of 50 units, unbound potency of 1.7 nM, efficacy 70 of units and sigmoidicity factor of 2.2, respectively. What would have been done if the three unbound concentration-response series did not match each other? In such a case, each series should be analyzed separately to yield a range of unbound potency EC_{u50} values. Depending on the range of EC_{u50} values, the effective concentration range may need to be expanded for estimation of dose in the dose nomogram which allows greater confidence in data from one particular species. Such a decision may therefore change the therapeutic effective concentration range.

Step 2b

In order to establish an effective therapeutic response, the hormonal response should reside within the 60–110 response interval. That is our goal with the biomarker. As shown in Figure 7.22 the corresponding free plasma concentration interval should range between 0.8 to 4 nM corresponding to a total plasma concentration range in man of 3.2 to 16 nM (based on a *fu* of 0.25). By multiplying the unbound concentration range with the human predicted unbound oral clearance shown in Figure 7.19 (*Cl/F* 5 to 6 L·min^{-1}), converting minutes to days (60·24), and using a molecular weight (Mw) of 275, a daily dose range of approximately 1.6 to 9.5 mg is obtained.

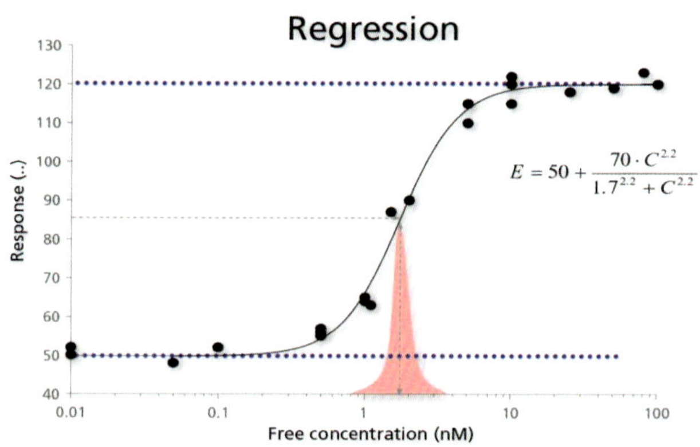

Figure 7.22
Pooled observed response *versus* free concentration (symbols) of all species superimposed on the model-predicted (solid line) response. The EC_{u50} (1.7 nM) value is indicated by a vertical arrow. The tentative free plasma concentration range 0.8 to 4 nM that needs to be established in order to get a response within the 60-110 response range is shown by the shaded red area.

Step 3

Combine the predicted human unbound oral clearance Cl_u/F of 6 L·min^{-1}with the predicted unbound plasma concentration 1.7 nM for estimation of the daily human dose.

$$Dose(mg\ per\ day) = \frac{CL_u}{F} \cdot 60(min) \cdot 24(h) \cdot C_e(nM) \cdot 10^{-9} \cdot Mw \cdot 1000 =$$
$$6 \cdot 60 \cdot 24 \cdot 1.7 \cdot 10^{-9} \cdot 275 \cdot 1000 \approx 4mg \qquad (7:21)$$

Step 4

Construct a dose nomogram (Table 7.9) by selecting an appropriate effective unbound concentration range and adjust the predicted human unbound oral clearance by a factor of 2 down and up (*i.e.*, -50 % down and +100 % up).

Predicted clearance	Effective unbound plasma concentration range (C_e) 0.8 nM	1.7 nM	4 nM
3 (L·min^{-1})	1 mg	2	5
6 (L·min^{-1})	2	4 mg	10
12 (L·min^{-1})	4	8	20 mg

Table 7.9
Dose-nomogram of Compound C based on predicted CL_u/F (-50 %, +100 %) and effective plasma concentration range ($C_{low} - C_{high}$)

The use of free concentrations is advantageous in at least three situations, namely

- when scaling animal data to man (if plasma protein binding is highly variable between species),
- when interpreting unconventional behaviour of total plasma concentrations, and
- following *iv* dosing of highly cleared drugs that have a narrow safety interval.

This Section has demonstrated inter-species scaling and pharmacodynamic modeling by means of fitting the sigmoid E_{max} model including a baseline parameter to unbound plasma concentration-response data at steady state and has shown how to scale oral clearance across species to man. The various concentration-response datasets obtained from mouse, rat and dog superimposed nicely when using free plasma concentrations. However, this may not always be so. Therefore, be aware that it may be necessary to select one or several animal species when you practice scaling. Using preselected response and concentration ranges together with the extrapolated human oral clearance, a human daily dose range has been estimated. The next step would be to design a clinical study including and testing the estimated, and perhaps an even wider, dosing range. The predicted unbound potency EC_{u50} (1–2 nM) value is an important parameter that is recommended as the target plasma concentration in future clinical studies.

Remember that the *target effect* should then ultimately govern the final dose selection for a larger patient cohort. In the aforementioned example, the E_{max} parameter was assumed to be the same for all species and man. However, in some situations there may be a need to adjust for species specific characteristics. One common example is differences in receptor densities and/or transduction efficiency, which may result in highly dissimilar response amplitudes and curve shapes between laboratory animals and man. *Affinity* and *intrinsic activity* are drug specific parameters, and what is observed in terms of E_{max} and EC_{50} may be confounded by the system. Whereas EC_{50} may be a perfectly valid measure in, for example, the rat, comparable pharmacodynamic efficacy may be obtained at EC_{15} or even EC_{100} in other species – depending on the relative impact of the target system in its particular, species-specific, physiological environment. It should further be appreciated that upon clinical testing in man, effects other than the desired may be encountered that constrain dose exploration and thus also the establishment of a full exposure-response curve. This will thereby pose a challenge for comparisons of potency and efficacy across species.

Table 7.10
Total concentration-response data of compound X in mouse, rat and dog

Mouse		Rat		Dog	
C (nM)	R (..)	C (nM)	R (..)	C (nM)	R (..)
1.04	50	-	45	.1	52
9.67	52	11	-	.49	48
49.9	56	17	55	5.05	57
111	64	38	63	10.6	65
189	-	51	87	22	90
510	110	64	-	-	-
837	115	162	115	104	122
2560	-	330	120	508	119
5300	-	820	118	987	120
9800	-	2700	123	-	-
-	-	3300	120	-	-

7.10 Principles of candidate drug selection for a first time in man study: Allometry, dose-nomogram and safety margin

7.10.1 Background

The aim of this Case Study is to practice your knowledge on the *evaluation process* of three potential drug candidates for a first time in man study. Imagine that there is an enormous unmet need for a drug within the therapeutic area of this particular chronic CNS indication. Therefore, your company is willing to sponsor any type of kinetic-dynamic study that can speed up the decision process and take the project promptly into man. This means that you should rank the three candidates based on *in vitro* data, *in vivo* binding and functional data after an acute dose, and *in vivo* safety data. Related aims are to demonstrate the difference between unbound and total concentrations and how they shift the concentration-response relationship, and how important it is to optimize the product of effective (target) pharmacological concentrations and clearance. This Case Study covers some factors of importance when one wants to predict a dose for a first time in man investigation based on sound pharmacokinetic and pharmacodynamic principles. Such factors include

- novelty of the agent (*e.g.*, is there structural precedence from which knowledge could be used to refine and assist modeling & predictions?)
- mechanism of action, knowledge of species-specificity of receptors
- its *in vivo* potency and predicted and observed receptor occupancy in man
- *in vitro* cell concentration-response relationships from animals and man
- concentration-target occupancy relationship *in vivo*
- *in vivo* concentration-biomarker and disease marker relationships in animals
- safety margin, ideally based on unbound concentrations
- pharmacokinetic-pharmacodynamic modelling.

Section 7.10.2 is a simple clinical dose prediction based on data from *in vitro* extrapolated human effective concentrations and clearance. Section 7.10.3 covers all available pharmacological data. Section 7.10.4 also includes *in vivo* safety data on *QT*-prolongation. Section 7.10.5 is aimed to show how a dose nomogram is constructed. Section 7.10.6 discusses design of some pivotal supporting preclinical studies. The data are thoroughly evaluated, the daily human dose is predicted, and the selected candidate is motivated and selected. A supplementary package of studies is suggested for which one has identified data to be either lacking or weak.

7.10.2 Combining extrapolated *in vitro* and *in vivo* data

In vitro and *in vivo* pharmacological data, plasma protein binding information, predicted human clearance and safety data are available for compounds X – Z (Table 7.11). Compounds A and C are agonists for two different targets and are part of an inlicensing discussion. The antagonist B, with its own specific target, has been developed inhouse. All three targets affect the same disease process.

Table 7.11
Selected compounds and their *in vitro* primate receptor binding, *in vivo* binding, *in vivo* response in dog, predicted total clinical concentrations, free fraction in human plasma, predicted human clearance and QT prolongation (dog model)

Compound class	*In vitro* IC_{50} or K_i (nM)	*In vivo* binding K_d (nM)	*In vivo* response	Predicted** clinical Conc (μM)	Predicted free clinical Conc (μM)	f_u	Pred i.v. *CL* ($L{\cdot}min^{-1}$)	QT dog (μM)
X. Mw 440	40–50	Tab. 7.16	–	3.1	~40 (nM)	0.013	0.14	See Tab. 5
Y. Mw 470	20–70	Tab. 7.17	–	>0.3	>10 (nM)	0.03–0.05	0.21	See Tab. 5
Z. Mw 318	9	–	Tab. 7.18	1.1–3.3	1–3**	0.9	8.4*	See Tab. 5

* Oral clearance based on rat and dog data after oral dosing
** Predicted from cell based *in vitro* assay data which are not included here

Evaluate the daily human dose based on clinical concentrations and predicted human clearance predicted from *in vitro* data (Columns 5 and 8, Table 7.11). Keep track of units. The estimated human bioavailability for compounds A and B are 75 and 85 %, respectively.

$$Dose(mg\ per\ day) = \frac{CL}{F}\cdot 24\cdot 60\cdot C(\mu M)\cdot 10^{-6}\cdot Mw\cdot 1000 = \left(\frac{CL}{F}\right)_u\cdot 24\cdot 60\cdot C_u(\mu M)\cdot 10^{-6}\cdot Mw\cdot 1000 \quad (7:22)$$

Where CL/F is the predicted total human oral clearance, 24·60 converts clearance (L/min) to cleared volume per day, C (μM) $\cdot 10^{-6}$ converts micromolar to molar units, and Mw converts moles to grams. The factor 1000 converts grams to milligram. Be consistent with multiplying unbound clearance $(Cl/F)_u$ with unbound effective concentration range C_u, or total clearance Cl/F with the total effective concentration range C.

Table 7.12
Selected compounds and their predicted effective total clinical concentrations, bioavailability, predicted human clearance and human dose

Compound class	Predicted** clinical Conc (μM)	*F*	Pred i.v. *CL* ($L \cdot min^{-1}$)	Dose (mg)
X. Mw 440	3.1	0.75	0.14	370
Y. Mw 470	>0.3	0.85	0.21	50
Z. Mw 318	1.1–3.3	-	8.4*	8.5g

* Oral clearance

** Predicted from other *in vitro* potency data which are not included here

The human dose prediction of compound X is

$$Dose\,(mg\;\; per\,day) = \frac{0.14}{0.75} \cdot 24 \cdot 60 \cdot 3.1 \cdot 10^{-6} \cdot 440 \cdot 1000 \approx 370\,mg \quad (7{:}23)$$

The human dose prediction of compound Y is

$$Dose\,(mg\;\; per\,day) = \frac{0.21}{0.85} \cdot 24 \cdot 60 \cdot 0.3 \cdot 10^{-6} \cdot 470 \cdot 1000 \approx\; > 50\,mg \quad (7{:}24)$$

Note that the predicted clinical concentration in Table 7.12 is given as >0.3 μM. Therefore, the dose prediction in Equation 7:24 will be greater than 50 mg. The human dose prediction of compound Z is

$$Dose\,(mg\;\; per\,day) = 8.4 \cdot 24 \cdot 60 \cdot 2.2 \cdot 10^{-6} \cdot 318 \cdot 1000 \approx 8500\,mg \quad (7{:}25)$$

Note that we use a midpoint concentration of 2.2 μM (range 1.1 to 3.3 μM) in Tables 7.11 and 7.12 for the dose prediction.

7.10.3 Assessment of all available *in vivo* data

The next step is to evaluate the daily human dose based on all available *in vivo* pharmacological data (Tables 7.16–7.18) and human clearance has already been predicted (Column 8, Table 7.11). Do not use safety (QTc) data at this stage. Aim for an occupancy of approximately 20–40 % of agonist A (Figure 7.23 and Table 7.12). Assume that the unbound brain concentration of agonist X is the same as in plasma.

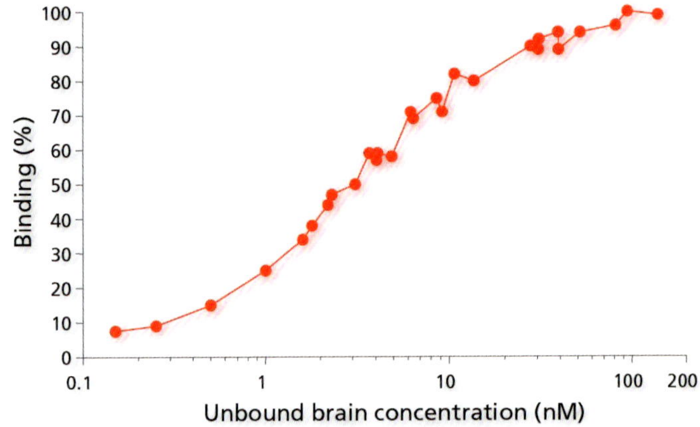

Figure 7.23
Receptor binding (%) *versus* log unbound brain concentration (nM) of agonist X in primates. Data taken from Table 7.16.

Figure 7.24 shows total and unbound plasma concentrations *versus* receptor binding (% occupancy) of compound Y. A receptor occupancy greater than 90 % is assumed to be required for an acceptable clinical effect. Therefore, you should give the effective plasma concentration as a range rather than a single point estimate.

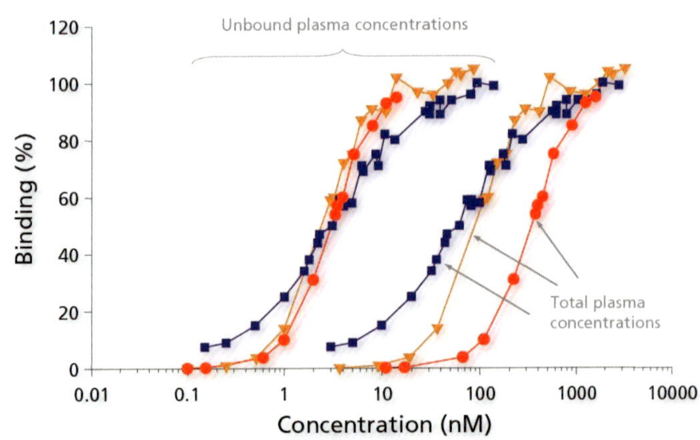

Figure 7.24
Receptor occupancy (%) *versus* log unbound (left) and total (right) plasma concentrations (nM) of the antagonist Y. Data taken from Table 7.17. Maximum receptor occupancy is needed for adequate pharmacological effect. Note the various shifts along the concentration axis due to differences in plasma protein binding.

The biomarker in the dog is a physiological marker which is intimately correlated to the disease in both animals and man (Figure 7.25). This *in vivo* biomarker can be viewed as an intermediate between the various *in vitro* assays and disease. Both the plasma concentration-time course, the baseline response-time course and the total response-time course is illustrated in Figure 7.25. This figure (right hand panel) also depicts total response plotted *versus* plasma concentration in time order, according to a clock-wise hysteresis. Consider what might be causing the apparent delay between plasma concentration and measured response. The aim is for 20–30 response units of pure *drug induced* response of the biomarker in dog for compound Z in order to establish an adequate clinical effect. A pure *drug induced* response means that the vehicle (baseline) response should to be subtracted from the total response curve (Table 7.18).

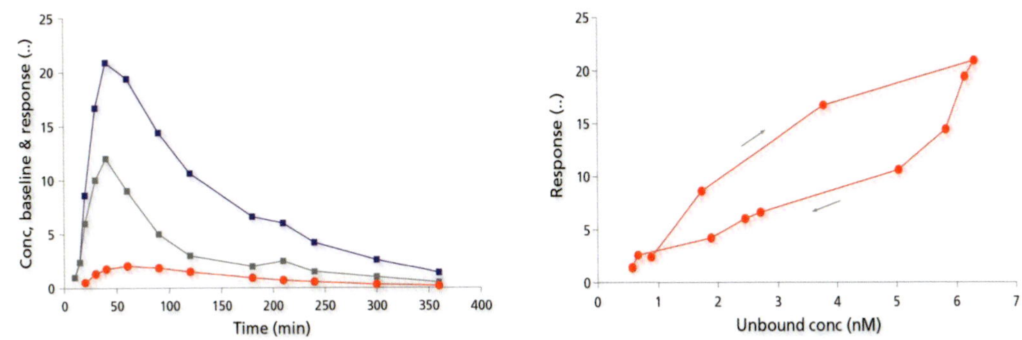

Figure 7.25
Left hand panel shows unbound concentration (nM, filled circles, bottom curve), baseline change (control, filled grey squares, middle curve) and total response (filled blue squares, top curve) versus time after an acute oral dose in the dog.
Right hand panel shows total response versus unbound plasma concentration (nM) plotted in time order (follow arrows). Data on compound Z are converted from μg ·L^{-1} (Table 7.18) to nM in the right hand panel above. Aim for a pure drug induced response of agonist C of 20–30 units.

Table 7.13
Selected compounds and their *in vitro* primate receptor binding, *in vivo* binding, effective *in vivo* concentrations in dog, free fraction in human plasma, bioavailability, predicted human clearance and estimated human daily dose

Compound class	*In vitro* IC_{50} or K_i (nM)	*In vivo* binding (nM)	*In vivo* response C_u (nM)	Free fraction man f_u	*F*	Pred i.v. *CL* (L·min^{-1})	Dose (mg)
X. Mw 440	40–50	1–2 (1.5)	–	0.013	0.75	0.14	15
Y. Mw 470	20–70	20–70 (45)	–	0.03–0.05	0.85	0.21	190
Z. Mw 318	9	–	>15 (20)	0.9	-	8.4*	85

* Oral clearance

The human dose prediction of compound X (Table 7.13) is

$$Dose\,(mg\ \ per\,day) = \frac{0.14}{0.75} \cdot 24 \cdot 60 \cdot \frac{1.5 \cdot 10^{-9}}{0.013} \cdot 440 \cdot 1000 \approx 15mg \qquad (7{:}26)$$

The human dose prediction of compound Y is

$$Dose\,(mg\ \ per\,day) = \frac{0.21}{0.85} \cdot 24 \cdot 60 \cdot \frac{45 \cdot 10^{-9}}{0.04} \cdot 470 \cdot 1000 \approx 190mg \qquad (7{:}27)$$

The original concentration-response data and baseline subtracted concentration-response data of compound Z are shown in Figure 7.26. The pure drug-induced concentration-response relationship appears to follow a straight line within the observed experimental range. The therapeutic goal was a response between 20–30 units. Assuming that extrapolation to a target response of 20–30 units can be made results in an unbound plasma concentration of about 15 to 25 nM with a mean of 20 nM. However, this is a linear extrapolation outside the observational range, which is a clear limitation. Therefore, one would like to study the response of compound C at two or more dose levels after repeated dosing to include coverage of a higher unbound concentration range in future studies. The dose for compound Z needed to establish an unbound plasma concentration of 20 nM is calculated as

$$Dose\,(mg\ \ per\,day) = 8.4 \cdot 24 \cdot 60 \cdot \frac{20 \cdot 10^{-9}}{0.9} \cdot 318 \cdot 1000 \approx 85mg \qquad (7{:}28)$$

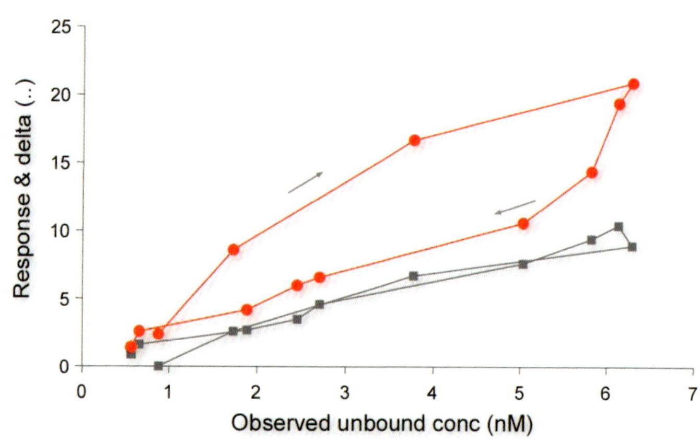

Figure 7.26
Unbound plasma equilibrium (steady-state) concentration (nM) *versus* total response (change from baseline %) of compound Z (agonist, filled red circles) and baseline-subtracted response (grey filled blue squares).

7.10.4 Assessment of safety margin based on *in vivo* data

Evaluate the daily human dose based on all available pharmacological data in Table 7.11 including safety data (Figure 7.28 and Table 7.17) and predicted human clearance (Column 8, Table 7.11). Figure 7.27 shows the unbound equilibrium (collapsed) concentration-response relationships which are based on data of Table 7.19. Rank the compounds by means of their C_{10}-to-C_u, namely the concentrations at a 10 % change from baseline in their QT intervals-to-the effective unbound plasma concentration for each compound in Section 7.10.3.

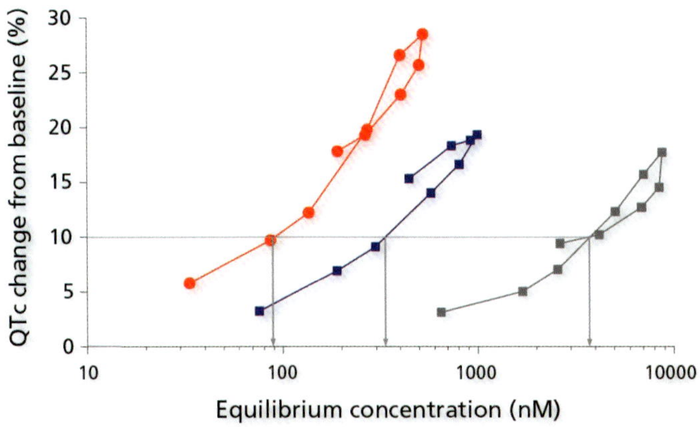

Figure 7.27
Unbound plasma equilibrium (steady-state) concentration (nM) *versus* QT-prolongation (change from baseline %) of compound X (agonist, filled red circles, left curve), Y (antagonist, filled blue squares, middle curve) and Z (agonist, filled grey circles); see also Table 5 (Appendix). The horizontal grey line corresponds to the 10 % cut-off QT response which should not be exceeded. The vertical arrows indicate the corresponding unbound equilibrium plasma concentrations for compounds X-Z. Note that these are approximate concentrations and you should therefore base the safety margin on a range of plasma concentrations (mean +/- 30 %).

Table 7.14 summarizes *in vitro* IC_{50} or K_i values, *in vivo* binding data, *in vivo* effective concentrations for the functional response, free fraction, concentration range for the occurrence of a 10 % change in the *QT*-response, the predicted human daily dose and the approximate safety margin.

Table 7.14
Selected compounds and their molecular weights, *in vitro* primate receptor binding, *in vivo* binding, effective *in vivo* concentrations from dog biomarker model, free fraction in human plasma, QT- prolongation (dog model), daily dose and safety margin

Compound class	*In vitro* IC_{50} or K_i (nM)	*In vivo* binding (nM)	*In vivo* response C_u (nM)	Free fractionman f_u	QT dog (nM)	Dose (mg)	Safety margin
X. Mw 440	40–50	1–2 (1.5)	–	0.013	60–120	15	~30
Y. Mw 470	20–70	20–70 (45)	–	0.03–0.05	200–400	190	<30
Z. Mw 318	9	–	>20	0.9	2500–5000	85	>100

Figure 7.28 is a comparison of binding data of *X* (*agonist*) and *Y* (*antagonist*) and functional response data of *Z* (*agonist*) with safety QTc data of all three compounds.

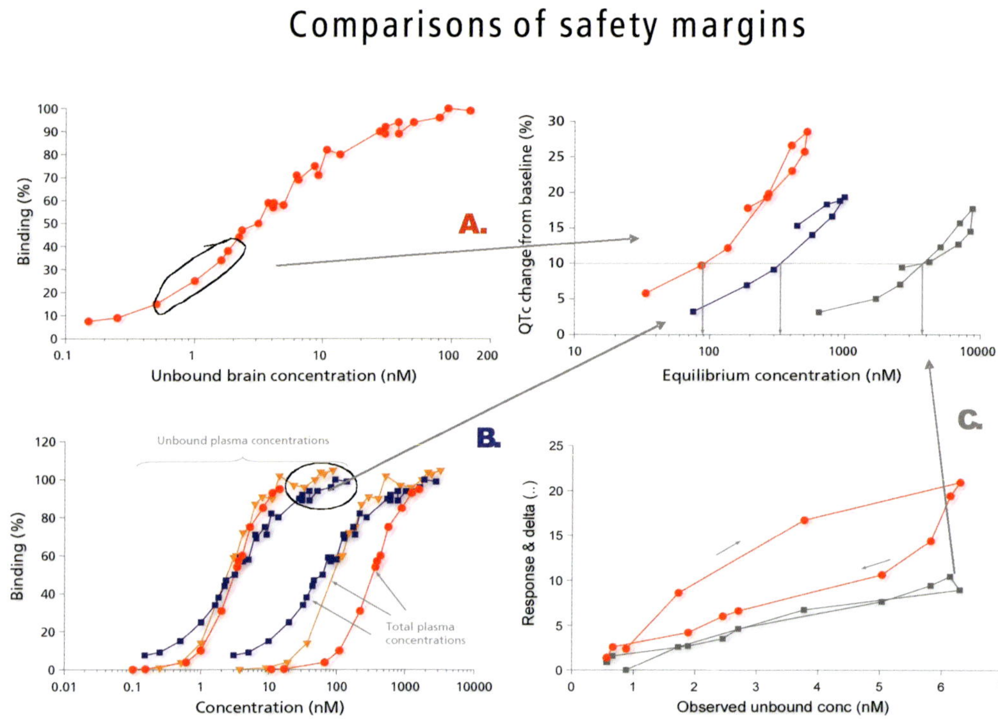

Figure 7.28
Upper left: Concentration-occupancy relationship of compound X.
Lower left: Concentration-occupancy relationship of compound Y in mice, rats and humans.
Lower right: Concentration-response relationship of compound Z.
Upper right: Concentration-response relationships of compounds X–Z with respect to QT changes in a dog model.

Remember that the assessment of safety margins is based on *steady-state plasma concentrations*. There is no information about the plasma concentration fluctuation (C_{max}/C_{min}) within a dosing interval in man. Ideally, this also needs to be factored into the analysis.

7.10.5 Construction of a dose nomogram

In an effort to visualize the consequences of uncertainties in both the plasma target concentration and predicted unbound clearance, the dose nomogram was constructed. The range of the unbound target plasma concentrations is based on the project estimated extent of variability in required exposure. Thus, the estimated effective concentration range may either

- cover everything from *in vitro* binding data (K_d, IC_{50} *etc*) to *in vivo* safety data (the concentration of a 10 % change in the QT interval from baseline) at high exposure
- be limited to a narrow range of the concentration-response relationship where also the selectivity of the compound is considered
- contain more or less arbitrarily selected concentrations, and then serve more as a starting point of distance to target predictions of the effective dose, or based on cost of goods constraints.

The expected uncertainty range in the predicted human clearance is either based on the observed variability of forerunner clearance, or a general appreciation of the distribution of clearance values in phase I studies, or otherwise predicted precision of human clearance obtained from preclinical data.

The dose nomogram is constructed as a visual means to present scaled data to decision makers, project groups or other scientists. For the decision maker it serves as a simple tool that conveys the combined effects of an unknown effective plasma concentration and predicted plasma clearance range. This creates a better understanding how either pharmacology (C_e range), DMPK (Cl range) or both are causing high or low uncertainty in the first dose to man estimates. The predicted human clearance of compounds X-Z are shown in Table 7.15. That prediction is coupled to an uncertainty of say -50 % to +100 %. Construct a dose-nomogram (See Table 7.15) for each of compound X, Y and Z. The predicted average daily human dose is derived from.

$$Dose\,(mg\;\;per\,day) = \frac{CL}{F} \cdot 24 \cdot 60 \cdot C_e(\,nM\,) \cdot 10^{-9} \cdot Mw \cdot 1000 \tag{7:29}$$

where CL/F, C_e (nM) and Mw are the total oral clearance, total (C_u/f_u) effective target concentration, and molecular weight, respectively. The '24·60' factor converts clearance from $L \cdot min^{-1}$ to $L \cdot day^{-1}$ and 10^{-9} converts nanomoles to moles. '1000' converts gram to milligram. Recall the units of each parameter and that the daily dose should be given in 'mg' units. Compound Z is used as an example of how the dose nomogram is constructed. Given an estimated total plasma clearance of 8.4 $L \cdot min^{-1}$ and three unbound plasma target concentrations of 20, 40 and 60 nM, the corresponding human doses are

$$Dose\,(mg\ \ per\,day) = \frac{CL}{F} \cdot 24 \cdot 60 \cdot \frac{C_u(\,nM\,) \cdot 10^{-9}}{f_u} \cdot Mw \cdot 1000 = \tag{7:29}$$

$$8.4 \cdot 24 \cdot 60 \cdot \frac{20 \cdot 10^{-9}}{0.9} \cdot 318 \cdot 1000 \approx 85mg$$

$$Dose\,(mg\ \ per\,day) = 8.4 \cdot 24 \cdot 60 \cdot \frac{40 \cdot 10^{-9}}{0.9} \cdot 318 \cdot 1000 \approx 170mg \tag{7:30}$$

$$Dose\,(mg\ \ per\,day) = 8.4 \cdot 24 \cdot 60 \cdot \frac{60 \cdot 10^{-9}}{0.9} \cdot 318 \cdot 1000 \approx 260mg \tag{7:31}$$

Table 7.15
Dose-nomogram of Compound *Z* based on predicted *CL/F* (-50 %, +100 %) and effective plasma concentration range (C_{low} – C_{high}) . C_{low} is taken from the target plasma concentration and C_{high} is a factor of 3 times C_{low}.

Predicted clearance	Effective plasma concentration range (C_e)		
	20 nM	40 nM	60 nM
4.2 L·min^{-1}	43 mg	85	130
8.4 L·min^{-1}	85	170 mg	260
16.8 L·min^{-1}	170	340	510 mg

7.10.6 Suggestions of pivotal preclinical studies

Since there is a huge unmet need for a drug in this therapeutic area, the company is willing to invest in further exploratory preclinical work for all three compounds. A couple of pivotal preclinical *in vivo* studies of compounds X, Y and Z have been proposed to be run in parallel with the first time in man study. These studies are in priority order.

I. Get a better understanding of the time-course of response at two or more dose levels coupled to both exposure and response steady-state. Repeated dosing at two or more dose levels and assessment of both exposure and biomarker/disease marker at steady-state for all three compounds in order to make a fair comparison with the same *in vivo* model. Remember, that two of the compounds are agonists and therefore functional adaptation (homeostatic feedback, tachyphylaxis, receptor down regulation, internalization and their time-courses) may be a potential source error.

II. If the company is to embark upon an lead optimization programme with follow up compounds, a better understanding of the link between *in vitro* binding and the *in vivo* biomarker *BM* response and outcome in the human (and animal) disease model is needed. For compounds Y and Z there seems to be some consistency between the *in vitro* K_i values and the *in vivo* binding K_d and *in vivo* target concentration.

III. Allometry based on more species. Compare *in vivo* measured clearance to that predicted from *in vitro* data for each species. An explanation is needed of how the human clearance was predicted from preclinical data. Estimate the precision of the method. Simulate the complete human concentration-time courses for all three compounds in order to assess fluctuation and how much greater C_{max} is than the average concentration.

Table 7.16
Unbound brain concentration (nM) and receptor occupancy (%) data of compound X in primates

Conc (nM)	Binding (%)	Conc (nM)	Binding (%)
0.25	9	8.6	75
0.15	7.5	6.4	69
1	25	6.2	71
0.5	15	3.7	59
4.9	58	13.6	80
1.8	38	4.1	59
9.2	71	93.6	100
4.05	57	30.4	89
2.31	47	138	99
1.6	34	27.6	90
3.1	50	80.7	96
2.2	44	39.1	89
38.9	94	51.2	94
10.7	82	30.6	92

Table 7.17
Total plasma concentration (nM) and receptor occupancy (%) data of compound Y in marmoset (free fraction in plasma f_u 2.7 %), Norwegian Blue cat (f_u 0.9 %) and man (f_u 5 %)

Monkey		Cat		Man	
Conc (nM)	Occupancy (%)	Conc (nM)	Occupancy (%)	Conc (nM)	Occupancy (%)
3.70	0.16	11	0.1	3	7.5
9.30	1	17	0.25	5	9
19	3.8	67	3.8	10	15
37	14	110	10	20	25
110	59	220	31	32	34
120	60	370	54	36	38
150	72	390	57	44	44
185	75	440	60	46	47
226	87	570	75	62	50
290	91	890	85	74	59
407	90	1230	93	81	57
520	102	1570	95	82	59
850	97			98	58
1220	96			124	71
1740	100			128	69
2110	104			172	75
2370	103			184	71
3190	105			214	82
				272	80
				552	90
				608	89
				612	92
				778	94
				782	89
				1024	94
				1614	96
				1872	100
				2760	99

Table 7.18
Time (min), unbound plasma concentration of agonist Z (µg ·L^{-1}), model predicted plasma concentration (µg ·L^{-1}, Mw 318), baseline response (..) and test compound response (..)

Time (min)	Measured conc (µg·L^{-1})	Pred conc (µg·L^{-1})	Baseline response (..)	Response (..)
10	-	-	1	1
15	0.28	-	2.4	2.4
20	0.55	0.53	6	8.6
30	1.2	1.34	10	16.7
40	2	1.79	12	20.9
60	1.95	2.07	9	19.4
90	1.85	1.88	5	14.4
120	1.6	1.52	3	10.6
180	0.86	0.92	2	6.6
210	0.78	0.71	2.5	6.0
240	0.6	0.55	1.5	4.2
300	0.21	0.32	1	2.6
360	0.18	0.19	0.5	1.4

Table 7.19
Compound, time (min), total plasma concentrations (µM) and original QTc (% change from baseline) response data. Free fractions of X, Y and Z of 0.01, 0.02 and 0.2, respectively, were measured *ex vivo* in dog plasma.

Cmpd	Time (min)	Conc (µM)	Effect (%)	Cmpd	Time (min)	Conc (µM)	Effect (%)	Cmpd	Time (min)	Conc (µM)	Effect (%)
X	0	0		Y	0	0		Z	0	0	-
	5.5	14.5	5.8		5.5	14	3.1		5.5	16.4	3.2
	10.5	18.8	9.7		10.5	18.8	5		10.5	19.7	6.9
	15.5	23.2	12.2		15.5	20.3	7		15.5	26.3	9.1
	17	-	-		17				17	-	-
	22.5	54.2	19.8		22.5	51.2	12.3		22.5	55.3	14
	27.5	64.8	26.6		27.5	49.6	15.7		27.5	60.5	16.6
	32.5	77.3	28.5		32.5	62.5	17.7		32.5	68.3	19.3
	34	-	-		34	-	-		34	-	-
	39.5	30.7	25.7		39.5	27.5	14.5		39.5	27	18.8
	44.5	21.8	23		44.5	18.6	12.7		44.5	19.3	18.3
	54.5	16.1	19.3		54.5	10	10.2		54.5	10.5	15.3
	64.5	13.8	17.8		64.5	7.9	9.4		64.5	8.7	-

Bibliography

Adams CP, Brantner VV (2010). Spending on new drug development. *Health Econ.* **17**: 130-141 (2010).

Andersson T, Weidolf L. (2008) Stereoselective disposition of proton pump inhibitors. *Clin Drug Investig.* 28(5):263-79. Review.

Ariens EJ. and Simonis AM. (1964) A molecular basis for drug action. *J. Pharm. Pharmacol.* 16: 137.

von Bahr C, Steiner E, Koike Y, Gabrielsson J. (1998) Time course of enzyme induction in humans: effect of pentobarbital on nortriptyline metabolism. *Clin Pharmacol Ther.* 64(1):18-26.

Blanchard J, Tozer TN, Rowland M. (1997) Pharmacokinetic perspectives on megadoses of ascorbic acid. *Am. J. Clin. Nutr.* 66: 1165.

Boxenbaum H (1980) Interspecies variation in liver weight, hepatic blood flow, and antipyrine intrinsic clearance: extrapolation of data to benzodiazepines and phenytoin. *J. Pharmacokinet. Biopharm.* 8:165-176,1980.

Bauer RJ, Dedrick RL, White ML, Murray MJ, Garovoy MR (1999). Population pharmacokinetics and pharmacodynamics of the anti-CD11a antibody hu1124 in human subjects with psoriasis. *J Pharmacokinet Biopharm.* 27(4):397-420.

Benet LZ, Hoener B-A (2002). Changes in plasma protein binding have little clinical relevance. *Clin. Pharmacol. Ther.* 7**0**:pp 115.

Brunton L L, Blumenthal D K, Murri N, Dandan R H, Knollmann B C (2011). Goodman & Gilman's The Pharmacological Basis of Therapeutics. 12th ed. New York: *McGraw-Hill.* ISBN 13:978-0-07-1624428.

Bull, H. G. G.-C. M., Andersson, S., Baginsky, W. F., Chan, H. K., Ellsworth, D. E., Miller, R. R., Stearns, R. A., Bakshi, R. K., Rasmusson, G. H., Tolman, R. L., Myers, R. W., Kozarich, J. W., and Harris, G. S. (1996) Mechanism-Based Inhibition of Human Steroid 5R-Reductase by Finasteride: Enzyme-Catalyzed Formation of NADP-Dihydrofinasteride, a Potent Bisubstrate Analog Inhibitor. *J. Am. Chem. Soc. 118*, 2359–2365.

Burris KD, Molski TF, Xu C, Ryan E, Tottori K, Kikuchi T, Yocca FD, Molinoff PBJ (2002) Aripiprazole, a novel antipsychotic, is a high-affinity partial agonist at human dopamine D2 receptors. *Pharmacol Exp Ther*. 302:381-389.

Calder WA III. (1981) Scaling of physiological processes in homeothermic animals. Ann. *Rev. Physiol.* 43: 301.

Cheung WK. and Levy G. (1989) Comparative pharmacokinetics of coumarin anticoagulants XLIX: Nonlinear tissue distribution of S-warfarin in rats. *J. Pharm. Sci*. 78: 541.

Copeland RA, Pompliano DL, Meek TD (2006). Drug-target residence time and its implications for lead optimization. *Nature Drug Reviews*. 1-10.

Copeland, RA (2011). Conformational adaptation in drug–target interactions and residence time. *Future Med. Chem*. 3(12), 1491–1501.

Danhof M, Alvan G, Dahl SG, Kuhlmann J, Paintaud G (2005). Mechanism-based pharmacokinetic-pharmacodynamic modeling – A New classification of biomarkers, *Pharm. Res.* **22**:1432-1437.

Dayneka NL, Garg V and Jusko W. (1993) Comparison of four basic models of indirect pharmacodynamic responses, *J. Pharmacokin. Biopharm.* **21**:457-478.

Diehl K-H *et al.* (2001) A Good practice guide to the administration of substances and removal of blood, including routes and volumes. *J. Appl. Toxicol.* 21: 15.

Farde L, Nordström AL, Wiesel FA, Pauli S, et al. (1992). Positron emission tomographic analysis of central D1 and D2 dopamine receptor occupancy in patients treated with classical neuroleptics and clozapine. Relation to extrapyramidal side effects. *Archives of General Psychiatry* **49**:538–544.

Fisher DM *et al.* (1985) Vecuronium kinetics and dynamics in anesthetized infants and children. *Clin. Pharmacol. Ther*. 37: 402.

Foster R. and Kreitzman L. (2004) Rhythms of Life: The Biological clocks that control the daily lives of every living thing. *Profile Books Ltd*., London.

Gabrielsson J, *et al.* (1985) Constant rate of infusion – Improvement of tests for teratogenicity and embryotoxicity. *Life Sciences.* 37: 2275.

Gabrielsson J, Jusko WJ, Alari L. (2000) Modeling of dose-response-time data: Four examples of estimating the turnover parameters and generating kinetic functions from response profiles. *Biopharm. Drug Dispos*. **21**:41-52.

Gabrielsson J and Weiner D. Pharmacokinetic-Pharmacodynamic Data Analysis: Concepts and Applications. 4th edition. *Swedish Pharmaceutical Press.* (2006) ISBN 13 978 91 9765 100 4.

Gabrielsson J, Green AR, vd Graaf PH. (2010) Optimizing *in vivo* pharmacology studies – Practical PKPD considerations. *J.Pharmacol.Toxicol.Meth*. 61: 146-156.

Hammarberg E, Svantesson GT, *et al.* (2000) Synthesis of novel 5-substituted 3-amino-3,4-dihydro-2*H*-1-benzopyran derivatives and their interactions with the 5-HT1A receptor. *J Med Chem* 43:2837–2850.

Holford NHG. *et al.* (1981) The effect of quinidine and its metabolites on the electrocardio-gram and systolic time intervals: Concentration-effect relationships. *Br. J. Clin. Pharmacol.* 11: 187.

In Vivo Pharmacology Training Group. (2002) The fall and rise of *in vivo* pharmacology. *Trends in Pharmacological Sciences* **23**:13-18.

Jonkers R *et al.* (1987) Beta-2-adrenoceptor-mediated hypokalemia and its abolishment by oxprenolol. *Clin. Pharmacol. Ther*. 42: 627.

Jusko WJ, Ebling W. (1995) Components of pkpd models. *J. Pharmacokin. Pharmacodyn.* 23:1-2.

Kapur, S., and Seeman, P. (2000) Antipsychotic agents differ in how fast they come off the dopamine D2 receptors. Implications for atypical antipsychotic action. *J. Psychiatry Neurosci. 25*, 161–166.

Kenakin, T (2006) A pharmacology primer: theory, applications, and methods (2nd Ed.) *Academic Press*. Chapter 3: Drug-Receptor Theory pp. 43-60.

Lappin G (2010) Microdosing: current and the future. *Bioanalysis* 2:509-517

Levy G. (1964) Relationship between elimination rate of drugs and rate of decline of their pharmacologic effects. *J Pharm Sci*. 53: 342.

Levy G. (1993) The case for preclinical pharmacodynamics. In: Integration of pharmacokinetics, pharmacodynamics, and toxicokinetics in rational drug development. Yacobi A, Shah VP, Skelly JP. *et al*, editors. *Plenum Press*, New York 7-13.

Levy G. (1994) Mechanism-based pharmacodynamic modeling. *Clin. Pharmacol. Ther.* **56**:356-358.

Levy G. (1994) Pharmacologic target-mediated drug disposition. *Clin. Pharmacol. Therap.* **56**:248-252.

Levy G. (1998) Predicting Effective Drug Concentrations for Individual Patients: Determination of Pharmacodynamic Variability. (1998) *Clin. Pharmacokinet.* **34**(4):323-333.

Levy RH. (1982) Time-dependent pharmacokinetics. *Pharmacol Ther.* **17**:383–397.

Levy RH. (1986) Time-dependent pharmacokinetics In: Pharmacokinetics: Theory and Methodology Edited by Rowland M, Tucker G. *Pergamon Press.*

Lovmar, M., Tenson, T., and Ehrenberg, M. (2004) Kinetics of macrolide action: The josamycin and erythromycin cases. *J. Biol. Chem. 279*, 53506–53515.

Maurer, TS, DeBartolo DB, Tess DA and Scott DO. (2005) Relationship between exposure and non-specific binding of thirty-three central nervous system drugs in mice. *Drug Metab Dispos.* **33**: 175-181.

Nagashima R, O'Reilly RA, Levy G. (1969) Kinetics of pharmacologic effects in man: the anticoagulant action of warfarin. *Clin. Pharmacol. Ther.* **10**:22-25.

Norberg, A., Gabrielsson, J., Jones, A.W., and Hahn, R. G. (2000).Within and between-subject variations in pharmacokinetic parameters of ethanol by analysis of breath, venous blood and urine. *Br. J. Clin. Pharmacol.* **49**, 399–408.

Nickolls SA, Fleck B, Hoare SRJ, and Maki RI (2006) Functional selectivity of melanocortin-4 receptor peptide and nonpeptide agonists: Evidence for ligand-specific conformational states. *J Pharmacology Exp Therap* 313: 1281-1288.

Oakes ND, Thalén P, Hultstrand T, Jacinto S, Camejo G, Wallin B, Ljung B. (2005) Tesaglitazar, a dual PPAR{alpha}/{gamma} agonist, ameliorates glucose and lipid intolerance in obese Zucker rats. Am *J Physiology Regul Integr Comp Physiol.* **289**: R938-R946.

Ollerstam A. *et al.* (2006) Pharmacokinetic-pharmacodynamic modeling of drug-induced effect on the QT interval in concious telemetered dogs. *J. Pharmacol. Toxicol. Meth*. 53: 174.

Oosterhuis B *et al.* (1986) Pharmacokinetic-pharmacodynamic modeling of terbutaline bronchodilation in asthma. *Clin. Pharmacol. Ther*. 40: 469.

Peck CC and Wechsler J (2002). Report of a workshop on confirmatory evidence to support a single clinical trial asabasis for new drug approval. Drug inf J. 36:517-534

Peletier LA, Gabrielsson J. (2009) Dynamics of target-mediated drug disposition. *Eur J Pharm Sci.* **38(5)**:445-64.

Piergies AA *et al.* (1987) Effect kinetics of N-acetylprocainamide-induced QT interval prolongation. *Clin. Pharmacol. Ther.* 42: 107.

Schwartz JB *et al.* (1989) Pharmacodynamic modeling of verapamil effects under steady-state and nonsteady-state conditions. *J. Pharmacol. Exp. Ther.* 251: 1032.

Scott JC *et al.* (1985) EEG quantitation of narcotic effect: the comparative pharmacodynamics of fentanyl and alfentanil. *Anesthesiology*. 62:234.

Segre G. (1968) Kinetics of interaction between drugs and biological systems. *Il Farmaco* **23**:906-18.

Sheiner LB, Stanski DR, Vozeh S, Miller RD and Ham J. (1979) Simultaneous modelling of pharmacokinetics and pharmacodynamics: Application to d-tubocurarine. *Clin. Pharmacol. Ther.* **25**:358-371.

Smith, DA, Di L and Kerns EH. (2010) The effect of plasma protein binding on in vivo efficacy: misconceptions in drug discovery. *Nature Reviews Drug Discovery*. **9**: 929-939.

Stanski DR *et al.* (1984) Pharmacodynamic modeling of thiopental anesthesia. *J. Pharmacokinet. Biopharm.* 12: 223.

Stone, S. R., and Hofsteenge, J. (1986) Kinetics of the inhibition of thrombin by hirudin. *Biochemistry 25*, 4622–4628.

Sällström B *et al.* (2005) A Pharmacodynamic turnover model capturing asymmetric circadian baselines of body temperature, heart rate and blood pressure: Challenges in terms of tolerance and animal-handling effects. *J.Pharmacokin.Pharmacodyn*. 32: 835.

Tabo M, Kimura K, Ito S. (2007) Monophasic action potential in anaesthetized guineapigs as a biomarker for prediction of liability for drug-induced delayed ventricular repolarization. *J. Pharmacol. Toxicol.Meth.* **55**:254-261.

Tummino PJ and Copeland RA (2008) Residence time of receptor-ligand complexes and its effect on biological function. *Biochemistry* 47: 5481-5492

Webb JA, Rostami-Hodjegan A, Abdul-Manap R,Hofmann U, MikusG, Kamali F. (2001) Contribution of dihydrocodeine and dihydromorphine to analgesia following dihydrocodeine administration in man: a PK-PD modelling analysis. *Br J Clin Pharmacol* **52**:35-43.

Whiting B *et al.* (1980) Quantitative analysis of the disopyramide concentration-response relationship. *Br. J. Clin. Pharmacol.* 9: 67.

Whiting B *et al.* (1981) Modeling theophylline response in individual patients with chronic bronchitis. *Br. J. Clin. Pharmacol.* 12: 481.

Wood, E. R., Truesdale, A. T., McDonald, O. B., Yuan, D., Hassell, A., Dickerson, S. H., Ellis, B., Pennisi, C., Horne, E., Lackey, K., Alligood, K. J., Rusnak, D. W., Gilmer, T. M., and Shewchuk, L. (2004) A unique structure for epidermal growth factor receptor bound to GW572016 (Lapatinib): Relationships among protein conformation, inhibitor off-rate, and receptor activity in tumor cells. *Cancer Res. 64*, 6652–6659.

Yang J, Jamei M, Yeo K, Rostami-Hodjegan A, Tucker G. (2007) Misuse of the Well-Stirred Model of Hepatic Drug Clearance. *Drug metabolism and Disposition*, **335**:501–502.

Symbols and their Definitions

Symbol	Definition or Explanation (units)
a, b, c, d	Allometric constants (arbitrary units)
A, B, C	Coefficients in the sum of exponentials (*e.g.*, $\mu mol \cdot L^{-1}$)
α, β, γ	Slope factors or exponents in the sum of exponentials (*e.g.*, h^{-1})
A_b	Amount in body (*e.g.*, μmol)
Amount	*e.g.*, μg, mg, g or μmol, mmol, mol
A_p	Amount in plasma (*e.g.*, μmol)
A_{ss}	Amount at steady state (*e.g.*, μmol)
AUC	Area Under the (zero moment curve) Curve (*e.g.*, $\mu mol \cdot h \cdot L^{-1}$)
B_{max}	Maximum binding concentration
bid	*bis in diem*, twice a day
BW	Body weight (usually g or kg)
C	Concentration (*e.g.*, $\mu mol \cdot L^{-1}$)
$C(t)$	Concentration at time t (*e.g.*, $\mu mol \cdot L^{-1}$)
C_A	Arterial concentration (*e.g.*, $\mu mol \cdot L^{-1}$)
C_{ave}	Average concentration (*e.g.*, $\mu mol \cdot L^{-1}$)
C_b	Blood concentration (*e.g.*, $\mu mol \cdot L^{-1}$)
C_e	Effect compartment concentration (*e.g.*, $\mu mol \cdot L^{-1}$) or effective plasma concentration
C_H	Hepatic concentration (*e.g.*, $\mu mol \cdot L^{-1}$)
C_{in}	Incoming concentration (*e.g.*, $\mu mol \cdot L^{-1}$)
C_M	Metabolite concentration (*e.g.*, $\mu mol \cdot L^{-1}$)
C_{max}	Maximum concentration (*e.g.*, $\mu mol \cdot L^{-1}$)
C_{out}	Outgoing blood concentration (*e.g.*, $\mu mol \cdot L^{-1}$)
C_p	Plasma concentration (*e.g.*, $\mu mol \cdot L^{-1}$)
C_{ss}	Concentration at steady state (*e.g.*, $\mu mol \cdot L^{-1}$)
C_t	Tissue concentration (*e.g.*, $\mu mol \cdot L^{-1}$)
C_u	Unbound concentration (*e.g.*, $\mu mol \cdot L^{-1}$)
C_{uss}	Unbound concentration at steady state (*e.g.*, $\mu mol \cdot L^{-1}$)
C_v	Venous concentration (*e.g.*, $\mu mol \cdot L^{-1}$)
C^0	Concentration at time zero (*e.g.*, $\mu mol \cdot L^{-1}$)
Cl	Clearance (*e.g.*, $L \cdot h^{-1}$)
Cl_H	Hepatic organ clearance (*e.g.*, $L \cdot h^{-1}$)
Cl_{int}	Intrinsic clearance (*e.g.*, $L \cdot h^{-1}$)
Cl_o	Oral clearance (*e.g.*, $L \cdot h^{-1}$)
Cl_R	Renal clearance (*e.g.*, $L \cdot h^{-1}$)
Cl_u	Unbound clearance (*e.g.*, $L \cdot h^{-1}$)
CNS	Central nervous system
CR	Controlled release
CSF	Cerebrospinal fluid
CV	Coefficient of variation also called relative standard deviation
Δ	Difference
D	Dose or daily or day
D_{ev}	Extravascular (*e.g.*, oral, subcutaneous) dose (*e.g.*, μmol)
D_{inf}	Infusion dose (*e.g.*, μmol)

D_{iv}	Intravenous dose (*e.g.*, μmol)
DMPK	Drug metabolism and pharmacokinetics
D_{po}	Oral dose (*e.g.*, μmol)
D_{sc}	Subcutaneous dose (*e.g.*, μmol)
ε	Intrinsic activity
E	Effect variable or Extraction ratio
E_0, R_0	Baseline effect (response)
E_H	Extraction ratio across the liver
E_{max}	Maximum drug induced effect (arbitrary units) or measure of efficacy
EC_{50}	Concentration at 50 % of maximal effect (*e.g.*, $\mu mol \cdot L^{-1}$) ; is a measure of potency
ED_{50}	Dose at 50 % of maximal effect (*e.g.*, μmol, or μmol/kg) ; is a poor measure of potency
EDA	Exploratory Data Analysis
EEG	Electroencephalogram
ev	Extravascular *e.g.*, oral or subcutaneous dosing
F	Bioavailability
F_a	Fraction of dose absorbed (*e.g.*, %)
f_e	Fraction of dose excreted via urine
F_H	Bioavailability across the liver
F_i	Fraction of binding to the ith site
F_{inf}	Infusion rate (*e.g.*, $\mu mol \cdot min^{-1}$ or $\mu mol \cdot h^{-1}$)
f_m	Fraction metabolized
f_u	Free fraction
FDA	Food and Drug Administration
Frct	Fraction
H	Hepatic
I	Intercept
I(C)	Inhibitory function, also known as 'drug mechanism function'
I_{max}	Maximum drug induced inhibition
IC_{50}	Concentration producing 50 % of maximal inhibition (*e.g.*, $\mu mol \cdot L^{-1}$)
ICH	International Conference on Harmonization
ID_{50}	Dose producing 50 % of maximal inhibition (*e.g.*, μmol, or μmol/kg); is a poor measure of potency
In	Input or incoming (arbitrary units)
in	Intraperitoneal (into *peritoneum*)
IRP	Indirect response
iv	Intravenous
K	Elimination rate constant (*e.g.*, h^{-1}) also called K_e
K_a	Absorption rate constant (t^{-1})
K_d	Apparent dissociation constant, ligand-receptor affinity (*e.g.*, h^{-1})
k_{e0}	Exit rate constant from effect compartment (*e.g.*, h^{-1})
k_{in}	Zero-order turnover rate for production of response
K_m	Michaelis-Menten constant (e.g., $\mu mol \cdot L^{-1}$)
k_{off}	Off rate (*e.g.*, h^{-1})
k_{on}	On rate (*e.g.*, $h^{-1} \cdot \mu mol^{-1} \cdot L^{-1}$)
k_{out}	Fractional turnover rate (*e.g.*, h^{-1})
K_p	Partition coefficient
[L]	Ligand concentration (*e.g.*, $\mu mol \cdot L^{-1}$)
λ_z	Terminal slope (*e.g.*, h^{-1})
LOQ	Limit of quantification or detection (*e.g.*, $\mu mol \cdot L^{-1}$)

m	Slope at EC_{50} on effect *versus* log-concentration curve
$MAPD$	Monophasic action potential duration
M_W	Molecular weight
MRT	Mean Residence Time (*e.g.*, h)
n	Sigmoidicity factor also called the Hill exponent
NCA	Non-Compartmental Analysis
P	Plasma or Precursor or Pool or Plasma protein
PD	Pharmacodynamics
PK	Pharmacokinetics
po	*per os*, oral
PPB	Plasma protein binding
Q	Flow or organ perfusion (*e.g.*, $L \cdot h^{-1}$)
Q_H	Hepatic blood flow (*e.g.*, $L \cdot h^{-1}$)
qd	*quaque diem*, daily, per day
R	Response, equivalent to pharmacological effect E
R_{in}	Rate of infusion (*e.g.*, $\mu mol \cdot h^{-1}$)
R_0	Dosing rate (*e.g.*, $\mu mol \cdot h^{-1}$) or baseline of response (arbitrary units)
R_{ss}	Steady state value of response (arbitrary units)
S	Slope factor (arbitrary units) or Substrate
$S(C)$	Stimulatory function also known as 'drug mechanism function'
sc	Subcutaneous
SD	Standard Deviation
SS	Steady state
t	Time (*e.g.*, s, min, h, d)
$t_{1/2}$	Half-life (*e.g.*, h)
$t_{1/2eq}$	Equilibration half-life (*e.g.*, h)
$t_{1/2z}$	Terminal half-life (*e.g.*, h)
t_d	Duration of response (*e.g.*, h)
T_{inf}	Infusion time (*e.g.*, h)
t_{lag}	Lag-time or time delay (*e.g.*, h)
t_{max}	Time to maximum concentration or effect (*e.g.*, h)
t_{ss} or T_{ss}	Time to steady state with respect to concentration or response (*e.g.*, h)
tid	*ter in diem*, three times a day
τ	Dosing interval (*e.g.*, h or days)
V	Volume or or volume of distribution (*e.g.*, L)
V_H	Volume of the liver (*e.g.*, L)
V_{max}	Maximum metabolic rate (e.g., $\mu mol \cdot h^{-1}$)
V_{ss}	Volume of distribution at steady state (*e.g.*, L)
V_u	Unbound volume of distribution (*e.g.*, L)
V_z	Volume of distribution based on the terminal phase (*e.g.*, L) also called $V_{d\beta}$ for a bi-exponential system
X	Independent variable or any parameter

Index